Art of Working Horses

Lynn R. Miller

Art of Working Horses
Lynn R. Miller
copyright @ 2016 Lynn R. Miller

All rights reserved including those of translation. This book, or parts thereof, may not be reproduced in any form without the written permission of the author or publisher. Neither the author or publisher, by publication of this material, ensure to anyone the use of such material against liability of any kind including infringement of any patent. Inquiries should be addressed to the publishers, Davila Art & Books and Small Farmer's Journal Inc.

Publisher
Davila Art & Books LLC.
in conjunction with Small Farmer's Journal
PO Box 1627 Sisters, Oregon 97759
541-549-2064
www.lynnrmiller.com

authored by Lynn R. Miller

First Edition Hard Cover, November 2016

Library of Congress Catalog Number
ISBN 978-1-885210-18-0

also by Lynn Miller
essays
    Why Farm
    Farmer Pirates & Dancing Cows
    Old Man Farming
poetry
    Thought Small
Fiction
    The Glass Horse
Non-fiction
    Ten Acres Enough: The Small Farm Dream is Possible
    The Work Horse Handbook
    Horses at Work (out of print)
    Training Workhorses / Training Teamsters
    Horsedrawn Plows & plowing
    Haying with Horses
    Horsedrawn Tillage Tools
    Starting Your Farm
    The Mower Book
    Complete Barn Book (out of print)

On the front cover; Ryan Foxley plowing on Littlefield Farm, photo by Joe D. Finnerty
On the back cover; Lynn Miller on the binder seat, photo by Kristi Gilman-Miller

This book is dedicated to the memory of

John W. Billington

A lifetime is not enough to repay you for the support and near on five decades of friendship you have shown us.

Hugo 'SJongers of France with four horses on a corn picker.

"Many people have sighed for the 'good old days' and regretted the 'passing of the horse'. But today, when only those who like horses own them, it is a far better time for horses."

~ C.W. Anderson

"There is something about riding down the street on a prancing horse that makes you feel like something, even when you ain't a thing."
~ Will Rogers

Leonardo DaVinci

Logging in NY, Dick Brown and Bill Grimm had to hook two teams together to get this big log up the "Old Dugout" skid road. Photo by Robert Mischka.

Mike Atkins competing at US Plow Championship, photo by the author.

William Castle of Shropshire, England, with hay rake in transport mode.

Preface

What is meant by working horses? The harnessed horse or mule out in front pulling or, in some cases, off to the side and maybe even pushing; connected. Near or further, tied to the governing human team-mate via lines, or ropes and/or voice - that might be a broad definition of what we mean by working horses. When we say working horses we are speaking of a particular job we may choose to do, to work horses - to employ the horse in harness to do certain labors.

Working horses, the life-long experience, has given me many of my best moments. This was made possible in the beginning by others 'handing off' to me what they knew and believed in. Now it's my turn. I want to hand off, to anyone who may have an interest, as complete a picture of true animal power as this one man has come to know. I wish to share how I might get willing and useful effort from an equine partner or partners. Along with that, I presume to share my understanding of the differing and unique ways of various other teamsters. There should be many views of, and recipes for, the actual working. As I see it, one way to accomplish as complete a presentation as possible is to bring together illuminating anecdotes along with technical explanations.

Back when I started working horses, all cars and trucks had carburetors, there were three television networks, no personal computers, no cell phones, no video games, gasoline was a quarter a gallon, coffee was a nickel (for all you could drink), no one was making new horse-drawn farm implements and free meals were served on airplanes. Back when I started out, looking into the realm of the craft of animal power was to gaze into a deep complex mysterious chasm. This mode of working had passed from modern man's ready knowledge. Conventional wisdom had it that no one took working draft animals as a serious option, not any longer. The ways to access the systems, to learn the craft, were gone or hidden from view. It was obvious that learning how to ask animals to willingly and effectively pull in harness was a tricky and varied business. Becoming proficient with this craft called for a subtle and nuanced way of working that required a lot of trial and error, a lot of research, a lot of patience. But that was way back then, half a century ago, not so true perhaps today. That isn't because the craft got easier to master. It's because the information became more accessible. We now have books, magazines, videos, and all manner of computer stuff which purport to show you the right way of doing this.

In keeping with the impatience of these times, I offer that too much of what is floating out there today about the teamster's craft is hooey, junk, blather, silliness, and often dangerous. Perhaps it is aggravated by t-shirted fellows sitting in suburban basements with soda and chips, pecking away on computer keyboards and espousing bizarre theories on any subject imaginable. They are 'experts', just ask them. Life on the ground floor may be passing them by but they are still going to try to stop traffic with their nonsense. Good folk wishing to learn about working horses deserve better for their investment and adventures.

The teamster's craft is finely nuanced. There is no way anyone can fully appreciate what is possible and all that is hazardous, those aspects which truly make of one a teamster, unless and until a full time, day after day, hour after hour, 'got to get this work done', month after month personal experience is under the belt. And that launch is best accomplished with experienced helpful folks in attendance.

True teamsters know each other by sight because the evidence of the experience comes through, it's in their carriage, their bearing, their 'ready patience'. And there is nothing about the teamster experience which makes of one a teacher. That's a whole 'nuther gambit. Yes, you can be a successful experienced teamster and a wonderfully effective teacher at the same time; Bob Olson (Colorado) and Kenny Russell (Mississippi) are two prime examples. But it is not requisite. There are thousands of great teamsters who aren't particularly good teachers, however most of them are sterling examples and phenomenal repositories of valuable knowledge. And further still, I am sure there are good teachers who have little or no hard-core working experience with mules and horses in harness. But as they are missing that aggragated experience, I have come to doubt their conclusions and their overall effectiveness. This observation and opinion is important here and now because today's "information is free" pollyannish attitude has already offered up too many newcomers to the slaughter of failed first efforts.

I believe that this way of working, the teamster's craft, has a vital future, but only if people interested in the system are given the best opportunities for success. At this late stage in my life I have decided that being diplomatic about critical differences of opinion on working horses does not serve the future. Here comes the core contradiction: Though this craft is complex and variable, applying sack fulls of theory, and training ritual, and drawing wild parallels to life's general difficulties make the process of learning too damned weird and drawn out. What this craft calls for is a solid understanding of the mechanics and a whole lot of sweat earned in the working. It should NOT be more difficult than that.

So I offer this book to any who might have the time for it. Here I hope you will find useful and accessible information wrapped in the wonderful contradictions of many examples.

Ray Drongesen mowing with King and Ruby on Lynn Miller's farm in 1975

"The horse moved like a dancer, which is not surprising. A horse is a beautiful animal, but it is perhaps most remarkable because he moves as if he always hears music."
~ Mark Helprin, A Winter's Tale

Preface	9	
Introduction	17	
Chapter one	23	Beginnings *Arrogance, Ignorance* *Harness and Harnessing*
Chapter Two	89	Progress *Where Were You When She learned That?* *Hoofcare*
Chapter Three	95	Mentors *How We Best Learn*
Chapter Four	125	Repetition *(Training)*
Chapter Five	133	Make Em Walk *(Plowing)*
Chapter Six	155	The Senses of the Horse *Levels of Communication*

Chapter Seven	169	Sharing the Load, Matching Time *Mowing*
Chapter Eight	179	Eye Sight and Rosie's Monster
Chapter Nine	189	Harness Essentials and Strip Pulling
Chapter Ten	193	Logging and Woods Work *Rube the Teacher (Bobsled)*
Chapter Eleven	209	Trust and the Ultimate Relationship *Lanas Bridle and Abe's Rebirth*
Chapter Twelve	219	Deep In the Mechanics *The Broken Strap*
Chapter Thirteen	223	Diagnose Comfort *Callie's Bad Day*
Chapter Fourteen	231	Best Horses *Polly and Anna's Lesson on the Buckrake*

Chapter Fifteen	239	What's Possible *Haying at Night - Feeding in the Dark*
Chapter Sixteen	243	Today and Tomorrow with Work Horses *Innovations / Applications*
Chapter Seventeen	295	Bits *A Nut on the Bridle and The Wrecking Service*
Chapter Eighteen	305	How Far Back? How Far Forward? *Mexican Beach and an Old Man's Reach*
Chapter Nineteen	311	How do You Know When a Horse is Broke? *John's Week and Leonard's Seventy*
Chapter Twenty	323	How Deep It All Goes *Bud and Dick*
Chapter Twentyone	327	Fine Tuning

Acknowledgements

Enormous thanks and gratitude to my patient and brilliant photographer wife, Kristi Gilman-Miller. To my editing daughter, Scout Gabrielle. To my brother Tony for doing chores. To my long suffering staff: A.J. Ferris, Shannon Berteau, Tasha Minke, and Eric Grutzmacher.

There are many photographers who have made this book possible. Some, because of the sluff of time, we do not remember, some we never knew, but several are unforgetable and central to this book.

My wonderful wife, Kristi Gilman-Miller, is the finest natural eye photographer I have ever known. William Castle of Shropshire, England, Joe D. Finnerty of Washington, Fuller Sheldon of North Dakota, Jean Christophe Grossettete of France, Robert Mischka, and so many more. Thank you all.

And to Ed Littlefield for believing in us.

All of you made this book happen. I am in your debt.

"Red Running" pastel by Lynn R. Miller

Art of Working Horses

Lynn R. Miller

Introduction

He was short and frail, at least eighty years old. Spoke in a whisper, kept his motions tight and in close, as if there was a chance an arm or leg would just fall off should he move too expansively. His neck seemed to have lost a lot of its mobility. You wouldn't expect him to be able to carry his own groceries out to the car but here he was driving a huge, jet-black team of Percheron horses perfectly fitted to an exceptionally well-built heavy leather harness. The team slowly, methodically pulled a flat stone boat sled on which old Norm stood riding. He held the two leather lines lightly but with perfect tension, careful not to lean back on them for balance. In fact, it was impossible to understand how he maintained his balance at all, standing as he did on that moving sled. I walked along entranced, didn't realize I was getting so close til I actually heard him whisper to those horses. "Steady." He said it quiet and natural. I could see the horses relax in their pull. I was twenty-eight years old, old enough to register that this observation would stay with me for a lifetime. I was watching craft and mastery while feeling the music it makes.

Almost four decades ago I wrote the *Work Horse Handbook*. I meant it to be an operator's manual. It's a pretty good book which has served many folks well. It deals with mechanics, the how-to's, with just a little on the why-fors. Back then I thought that was all anyone would need, so I didn't get into the myriad intangible aspects that make of this way of working a craft.

Many years after the *Work Horse Handbook* I wrote *Training Workhorses/ Training Teamsters*, an ambitious volume that set as its goal a presentation of information folks could use to train themselves and their horses. An extension, if you will, of the operator's manual. This is another book that tries to tell people how to work animals, and how to train animals to accept and excel at the work.

Ray Drongesen and Lightning the mule by Nancy Roberts

Lynn, Bob, Bud and Ray in a 'Pepsi Generation' ad photo from 1975

There have been several other volumes in what we refer to as the *Work Horse Library* but long ago I realized that none of them adequately acknowledges the magic which is possible with this craft. A close working communion with draft animals is by its nature mysterious, constantly serving up surprises that even the old masters frequently didn't and don't see coming.

When I first started out, as a young man trying to work horses, I was an embarrasing novice. I had a friend who was a masterful working teamster, an uncommon farmer living a common and accessible life centered around his draft horses. I was lucky enough to have my first farm near his and to have him as both a mentor with the horses and as a farming partner. Ray Drongesen was in his sixties then, as I am now, and he had been forced to retire from his mill work early because of a heart condition. It was expected that he would take it easy on his disability. Instead he amped up his part time hobby of a team of working horses to full-time with four then six head, including a working stallion, renting idle tracks of land inside his small town and putting up grain and hay. When I came along we shared work and I gave him a piece of the harvest on my seventy-seven acres in trade. As our work-day worlds drew more and more local media attention, we both had folks drawn to us because they thought they wanted to do this workhorse stuff as well. Out of that environment sprang the idea for the *Small Farmer's Journal*, a quarterly publication that created an agrarian-based community for far-flung horse farmers and loggers.

Ray and I came up with the notion of doing a Work Horse Workshop or two on my farm and that evolved to annual Work Horse Auctions which included instruction venues.

By this time I had been working horses for only a few years, but I was totally dependent on them to get my farming done. For this reason many of the lessons learned and the key elements of the mechanics were sharp in my mind. Perhaps that's why when I first heard Ray say it, I was confused, even mystified. I didn't have a clue what he meant. We were working with some students and he said "The longer I work horses the less I know and the easier it gets." Now, with the passage of time, it makes perfect sense to me. But back then…

Back then I was wrapped up in trying to find better ways to teach people how to work horses, so prescriptions and check lists of just exactly what a person should do when hitching or unhitching, when driving and adjusting for implement efficiency, all of these things where vitally important to me. And I thought they should be equally important to my students. Ray's statement flew in the face of that and seemed to set all of our work as instructors topsy turvey. On the surface it seemed he was saying that all those details, all of that technique, all of the sort of stuff that we think goes into mastery of a craft - that all of that stuff was not that important, perhaps even less than important.

I stored it away as an intriguing piece of this man who I revered. I guess, early on I figured that some of that strangeness was what made it all magic and that I might not ever know what he was talking about. But long ago, it started to make perfect sense to me and even help me to get out of my own way. There I go, sounding all vague and strange, just like Ray only not nearly so wise. What I mean by "getting out of my way" is that frequently I overthink my work, in advance and during. I analyze and speculate and fidget and worry things along when if I just quit thinking so much about it and went along with the working, things would move so much smoother, and the horses would become mellower with each step.

In fact, that now it is a central tenant of my philosophy of working animals. It's not just an observation, it's a goal. The magic flows when we become inseparable from that working partnership, that communion of effort. And you don't get there from a check list. You get there by going there. You get there by hours and hours and days and days and weeks and weeks of actually working the animals until you find yourself working alongside them and finally working with them.

Bulldog Fraser of Noxon, Montana, spreading fertilizer in his portable 'shade'.

Horses or mules, donkeys or ponies, regardless the size or variety: this writing is about the mysterious craft that would have us successfully calling upon those grand beasts to wear a harness and pull, with precision, some vehicle or implement for us - the overall objective of which is to get some work done, cover some ground, move to another place.

There are invisible structural elements that go towards the success with working horses. You can't see all the things that build towards the balance and the quiet willing companionable work. And in the beginning perhaps you don't need to see those things, perhaps it's good enough that you understand the mechanics, the suitable precautions, the limits; and that you adhere to them. After all, none of what is possible will ever come to pass if you and/or your horses are afraid, are uncertain. And it won't occur if you get hurt.

But to contradict myself, please know this; you can be successful at working horses without ever being artful. It can be just another way of working, matter-of-fact and precisely disciplined. But if that is the case you may have cheated yourself and your working partners out of the entrancing magic that gives us the art. To be artful with the teamster's craft requires a rare combination of human qualities; absolute patience mixed with determined gratitude and blended with the good eye.

There is what some might see as a curious construct with this book. Years of chewing my ideas and thoughts on the subject, I landed firmly on the desire to put together this volume as a four-dimensional conversation. I wanted to share my stories and the stories of others, allowing for disagreement, and reaching back in time to tie it all together with narrative elastic offering a chance at immersion. When I used to do public speaking my wife Kristi, made the observation that I did a better job of communicating when I spoke off the cuff and without notes; better than when I carefully constructed an essay for oral delivery. That frequently came off dry. I took her advise and found she was right. I've tried to build this book in the same informal way, trusting that it will draw you in to discover information in a natural rythmn. I am reinforced in that objective because I see it as paralleling one of the strongest attractions of the teamster's craft; it allows us to be in a natural rythmn. I may never know if it worked, but I trust the effort.

This book then is to be a sharing of the stories, maps, recipes, snapshots, sketches, warnings, exaltations - all of it from the middle of the art of working horses.

'If time were the wicked sheriff in a horse opera, I'd pay for riding lessons and take his gun away.' - W. H. Auden

'I don't even like old cars. I'd rather have a goddam horse. A horse is at least human, for God's sake.' - J. D. Salinger

Ed Joseph harnessing a horse in 1999. Photo by Kristi Gilman-Miller.

Chapter One

Beginnings

Arrogance, Ignorance, and the Mechanics of the Working Harness

It's been a long time ago now, from 50 to 55 years or so, back when I felt myself drawn to each and every image I ever came across of horses or mules working in harness. Didn't know why. I was a city kid with no immediate background in the stuff. But the attraction was very strong and central to an overarching dream of someday having my own old-fashioned, general farm. Back in the fifties and early sixties, in urban centers, the conventional wisdom had it that agriculture had grown up for good and all. The industrialization of farming with its concomitant chemistry and heavy metal disease (big machinery) was, back then, already fashioning a base for today's genetic engineering and cyber nonsense. Way back then, anyone who expressed an interest, let alone a preference, for the old farm was branded as "backwards" and "sentimental."

I was a kid and had no ammo to argue for my dreams. So I didn't. Instead I just allowed myself to be carried away privately, secretly, each and every time an image tugged at me. Those

A student on the plow handles at Russell's Mississippi Workhorse Workshop.

images, big barns with open hay mow doors, rows of pruned fruit trees, a line of Jersey cows coming in at dusk to be milked, chickens pecking at proffered grains, sheep clustered 'neath a tree for shade, grain fields wafting in the breeze, freshly turned earth steaming midmorning, old tractors popping along, all of it, any of it, took me some place I didn't know, yet... it was some place quite familiar. It was like I carried around with me some sort of genetic memory of farming from days gone by. Sights would take me there, smells also and sometimes even sounds. And the one category of those things that had the strongest draw for me was any view of working horses or mules. They didn't have to be moving, they could just stand there patiently, wearing the harness like a suit of utilitarian armour. Sure, the images of four horses at a gallop pulling a western stage coach, one guy on the high seat handling the lines, that was exciting. But, with time, that didn't thrill me as much as the view of two horses pulling a walking plow with precision, or four horses walking side by side pulling a harrow across the plowed ground, or a big gelding skidding a log through the forest. And in each and every case someone, a teamster, was in careful attendance. A little later in life when I took up cowboying, I had a tactile sense of what it meant to sit astride a horse and subtly get it to move the way I wanted. (I also had plenty of experiences when it didn't work and the errant horse beat me up.) But in those early years I couldn't quite figure out how a mere mortal, back behind the harnessed animals with just some leather straps in hand, could control the animals speed and direction. It didn't bother me much, because back then I honestly didn't think I'd ever have a chance to actually try to work horses myself.

Early seventies, I was managing a sheep and cattle operation in coastal Oregon on a farm that had a lot of christmas trees, not the carefully cultivated type, rather the volunteer Douglas

We enjoy the romance of horses galloping across the landscape but horses in harness running away in complete terror; we don't like that. It is an outcome which deserves our best effort to avoid.

firs that grew all over the hill pastures. They were uniform and of the perfect size. The owners wanted them removed and I got the idea that I could sell them to the holiday trade. But there was a challenge. In the winter time it rained constantly and those hillsides were mud waiting to happen. Neither the little Ford 8N tractor nor my pickup could make it up those hillside roads to retrieve the trees. So, for reasons I still don't understand, I got it in my skull that if I had a team of draft horses I could go up and down those hills with loads of christmas trees. Long story shortened, my then boss loaned me the money and I went to a liquidation auction and bought two Belgian mares, both well broke. The seller threw in an old harness and collars. I was set, or at least I thought I was.

At that point in my life, I was in my twenties, I was riding an excellent cutting horse to separate cows for breeding artificially. Doing that work spring and fall and pretty regular. I figured if I could do that I could probably handle something as simple as driving a team. I had paid attention and thought I knew the system. It was Thanksgiving weekend. I had butchered one of my own geese for supper (it was in the oven) and was feeling particularly handy so I decided to harness up the girls and hitch them to a flat sled. I had no help and figured I didn't need any. Got the horses outfitted and led them out to stand in front of the double-tree (no tongue on the sled). Then I rigged up the driving lines how I had seen it done and hooked the trace chains to the single trees. The horses stood perfectly still the whole time. Then I stepped onto the sled and slapped the lines while hollering "Hyah," like some idiot on television. The mares jumped ahead, at first hesitant, but with enough surprise to throw me off the sled. I dropped the lines and the mares broke into a run.

I ran after them yelling whoa, as if I had some reason to believe that would accomplish anything. It was a small three acre triangular pasture with steel posts and woven wire fencing topped with a single strand of barb wire. They ran towards a corner trying to slow down but every time they did, the sled and evener would hit them in the heels and off they'd go again. I could feel a sharp pulsing pressure of fear and terror not just inside of me but in the air as well. They turned, but the dynamics of the trailing sled pushed Queenie into the fence and then through it, so that for fifty feet they straddled the fence popping and bending posts and getting tangled in the wire. Finally Goldie was tripped by the wire tangle around her pasterns, she went down and rolled to her back; Queenie stopped and stood shaking. I ran to them. The sight of the two of them that way drove me to tears. Still stupid, I methodically unwrapped the fence wire from Goldie. It took some time, luckily I had a pair of pliers with me to cut the tighter wraps. Goldie laid perfectly still, completely drained of strength, in a heightened state of shock, and watched me working. Queenie stood quietly. Must have taken over a half hour to get everything undone. I led the mares to their stalls in the barn and pulled the busted harness off them.

Had I been duped? Had I been sold a damaged, unreliable pair of horses? Something, right then, told me no. Maybe it was the incredible patience these two had to not fight the tangle and stand allowing me to free them from the wire. These were good mares. The problem was me. The problem was that I had no idea how uninformed I was. I remember asking myself, over and over again, "What have I done, what have I done?" I could not figure out what went wrong. Now of course I can look back and see all the problems, but back then I was clueless.

I had heard about a local man, Howard Steele, who had a team of Belgians he used in pulling matches. He was a bit of a legend in those parts. I learned much later it was a positive, earned reputation because Howard was enormously successful at pulling and he did it quietly, purposefully and without any outside help, or swampers as they are called (people who held the horses and did the actual hooking to the eveners). He hooked his horses, Rube and Champ, himself. He spoke softly to them, he assured them, and set as his insistence that they remain calm.

I went to see him at his small farm. I introduced myself and told him, in as much detail as I could recall and understand, just what had happened. I asked him if he could help me. He drew a diagram and told me to go home and build a tongue for the sled PRECISELY as he drew it.

Later, when Howard came to my place he said, "I want you to do exactly what you did before. I want to watch you harness and hitch those horses."

"But," I interupted, "they'll just run again..."

"Just do what I ask."

So I set out to do just that. I had the team tied to a hitching rail and harnessed and was rigging the driving lines when he stopped me.

"Stop right there, I can't take any more of this. Just stop right there. What on earth are you doing?" he said in an angry voice. He came forward and started unsnapping, twisting, flipping and moving harness parts and lines.

I asked, "Why are you doing that?"

He stopped and turned to glare at me, "You've got no right to have horses like this. YOU will never make a teamster. I'm not doing this for you. These horses need to be 'unwound' from that wreck. I'm going to fix this mess and drive them. You can watch if you want but don't bother me with your stupid questions. Just stay out of our way."

Struck dumb, or should I say dumber, I lit a cigarette and stood up against the shed wall to watch, feeling humiliated and a little angry. Within minutes Howard had rearranged the harness and lines. Then he untied the team from the rail and walked behind with the lines. Speaking firmly but quietly he had them step ahead and with two attempts got them to drive over the new tongue in position to hitch. Lines off to the side, he went forward and hooked the new neckyoke then walked back, picking up the lines, and quietly hooked the trace chains to the single trees. The mares stood still but they were starting to twitch. It was easy to see they were nervous. Howard stepped onto the sled and, holding the lines still and with no slack, he spoke to them to go. They stood still. He pulled the lines back just a little then gave some slack and used a more forceful and surprising tone to his voice. The mares stepped ahead and when they heard the sled runners on the gravel they dropped their hips as though to lunge forward. My heart jumped. Howard stayed calm, standing on the sled, and pulled back on the lines just a little while quietly saying "steady". The mares danced a couple of steps and then settled into a perfect walk. The man, team and sled went off down the farm road while I trotted behind with a mixture of thrill and dread.

Everything went smooth, smooth enough for me to ask Howard, "Can I drive them now?" He glared over his shoulder at me and shook his head no. He drove the mares for about a quarter of a mile on the sled, returning to where he had hitched. Stepping off the sled, lines in hand, he set to unhitching while he said, "I'm going to unhitch this team and then this is what you are going to do. You are going to promise me that you never again try to hook these good mares. Anybody knows that cigarette smells and horses do not mix. You are going to sell this team and this harness and thank your lucky stars you or they aren't dead because of your arrogance and stupidity."

LESSONS

In retrospect I understand now the core of Howard's lesson, tough as it was to swallow. I did not make that promise to him and to my credit I went far and wide in search of instruction and successfully put those good mares to years of work. As for that first time hitching I believe, in my stupidity and ignorance, that I had the team lines backwards, the hames hooked so loose that they popped out of the collar groove which had the bottom hame strap choking the horses, the quarter straps snapped into the trace chains, and heaven knows what other mess-ups. The sled runners were steel and, without a tongue to work as a backing and breaking system, they ran easily on the wet pasture grass causing the sled to slide up into the back legs of the horses when they would try to slow or stop. Slapping the lines and yelling is no way to start a team of horses, let alone a working relationship.

Smells are associative. It was simplistic to say that a smoker couldn't be a teamster. I quit smoking in 1985. I worked horses successfully before and after that without noticing any adverse reaction from the horses. My mentor, the exceptional teamster Ray Drongesen, was never without a cigar in his mouth. But this is not an attempt to take anything away from the excellent Howard Steele. He had and has a system, a set of rules and rituals that served him well. He stuck to them, all of them, and they made him what he is.

I might add that I too stuck to my 'guns' and never gave up on my desire to learn the craft and earn the right to be called a teamster. It took a long time and many knocks and it remains one of my proudest achievements. Learning to work horses is akin to learning a new language or how to master a musical instrument, it requires real dedication and a long committment. Learning to work horses shapes a person, permanently.

Richard Douglas, horsefarmer of New York.

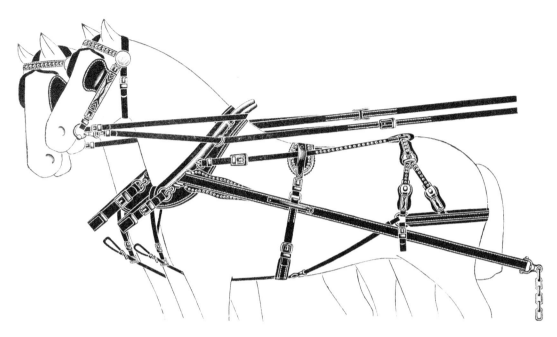

A western-style two-strap brichen farm harness with conventional tugs. No nose bands on bridles. A rudimentary harness suitable for most any farm and wagon work.

HARNESS

The mechanics of the North American system of working horses and mules in harness developed over approximately three hundred years and were guided in the beginning by a rich tradition of craft-based tradesmen who enjoyed 'freedom' of artistic circumstance when first coming to this continent. At the time the first trans-atlantic Europeans came to the new world they had to make do with what skills and tools they brought with them, there was no 'local' infrastructure of support in what was then seen as a wilderness. This reality somewhat freed the leather workers from the rigid bounds of a thousand years of tradition and caste, tradition going back to the first civilizations of the Middle East and eventually the Roman empire. As horses and mules became more readily available in the North American colonies, local tradesmen developed their own variations on the harness systems from the old countries. The basic framework remained the same, but many unique distinctions arose.

For each draft animal employed, the torso-encompassing harness system featured an engineered web of straps attached around the neck and from the chest and/or shoulders, translating the pushing of the animals to a pulling action. Expectations and successes had the approach quickly evolve to a modular system, allowing that multiples of animals be hitched together for greater force and applicability. The harness involved some controlling restraint at the animal's head, usually a 'bit' in the mouth, but also, early on, it featured the presump-

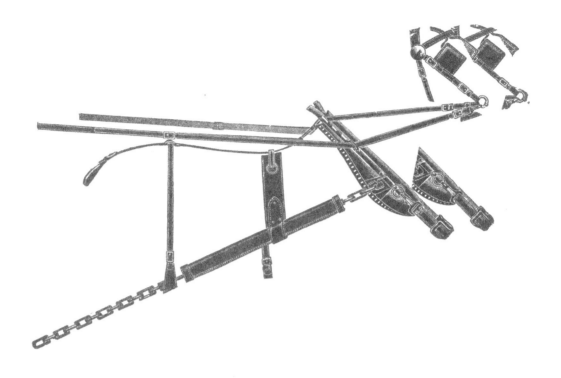

This is an interesting image as it allows you to see how the shape of a harness on a team should fit as if it is a definition of the form of the animals. In this case the harness is a chain tug, Southern-style logging harness with no brichen, only cruppers around the tails.

tion that sometimes the animal might be led by the head with a lead shank and a 'halter'. The modern system employs long, segmented driving 'lines' once made of rope or leather and today more commonly made of either leather or synthetic materials such as *Beta* (a material composed of petroleum-based synthetic substances). The animals were and are outfitted with either a wide strap-like breast collar or a stuffed encircling leather collar both of which they push against. From the ends of the breast collar run, backwards along the animal, heavy straps (or chains) called tugs (or traces) which hook to a 'singletree'. Or, in the case of the stuffed leather collar, an offsetting matched pair of tube-steel or reinforced wood "hames" (rib-like in form) are affixed into the collar groove, these feature a bolt-assembly, at approximately one third from the bottom of the collar, to which are fastened two tugs or traces (one to a side) and each running parallel back along the animal and fastening to the apparatus which in turn fastens to that which is being pulled.

As this harness system evolved, from region to region, increasing attention was put to the question of efficiency of draft - or said another way - the ease with which the draft animals might pull the load. Many saw, early on, that this was attained only if the comfort of the

A fancy, heavy coaching harness.

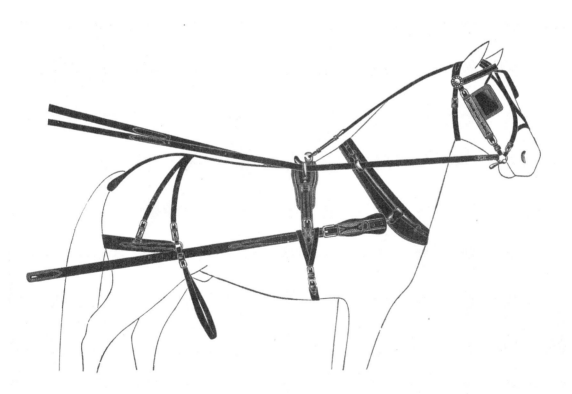

A straight forward, plain, hame and collar-style buggy harness of heavier contruction and design.

Bobby Jones' three fine Belgian horses hitched to a Pioneer sulky plow and competing in the US Plow Championship in Dalton, Ohio. With western brichen-style harness, when well-fitted as you see here, horses work comfortably. Photo by author.

working animal was fully considered. But also, and included, there were laws of physics to keep to the forefront of design. For example; if the load to be drawn was to be drug along the ground's surface, it became advantageous to reduce the friction by allowing that the forward motion also accomplished a slight lifting of the front of the load. (In this way the load became easier to pull.) It was found that combining a comfortable and well-fit harness with the right leverage resulted in improved or even maximum draft efficiency.

When two or more animals are employed to pull a wheeled vehicle or implement it is often vital to utilize a suitable "backing and braking" system. The Western Basket-brichen harness design employs a Y - shaped system which runs forward under the belly, from the two ends of the brichen through two adjustable quarter straps. Those quarter straps then fasten to the back end of a pole strap. The front end of that pole strap fastens to one end of a neck yoke. This yoke affixes to the front end of the pole or tongue at its center. (See page 87) With

It takes an experienced eye to "see" the fit of harness when the horses are standing unhitched. This Amish team at the 2016 Horse Progress Days is perfectly outfitted in a synthetic harness. Photo by Jerry Hunter

this system, when the two animals stop, the forward motion of the implement or vehicle is transferred along the tongue and held back by its arrangement with the pole and breast strap assemblies. Those assemblies are fastened to the two ends of the neck yoke.

One notable group of variations on this structural dynamic comes with the side-backer and D ring styles of harness, wherein straps or tugs are coming forward from the brichen ends alongside the animal to jockey-yokes which function, off the ends of the neckyoke, like a second evener only for the backing and braking. (See diagrams and note Sidebacker versus D ring harness on pages 48-51).

Over the last three decades there have been many advances within the arena of true horse-power. With the harness perhaps the most dramatic innovation, one which many, if not all, teamsters employ to advantage, has been the replacement of leather with synthetic materials. A complete single leather harness for a full-sized draft horse might weigh from a low of fourty pounds to nearly eighty-five pounds. Lifting that harness up five - six feet in the air to pass over the back of a waiting horse takes a tall, healthy person's physical strength. A synthetic harness weighing from 20 to 35 pounds becomes a real convenience for many people.

Not all synthetic harnesses are equal. With time, new innovations in materials and construction have resulted in vast improvements in synthetic harness construction. Whether a person is getting started with leather or synthetics, it is important to have a knowledgeable individual assist in determining the quality and suitability of the harness being considered. Sometimes common sense may be that guide. If a team of horses have used a particular harness to comfortable advantage for a period of time long enough to safely say it fits, a premium should be paid to keeping that harness with those animals as they change ownership.

(A note of caution: over the last several decades extremely cheap harness of dubious strength and configuration has been imported from India, Pakistan and other parts of Asia. While appearing suitable, and often decorated with ornamental hardware, the straps are frequently of random lengths and the leather of questionable strength. This author has yet to see one of these harnesses of a recommendable quality.)

HARNESS FIT

The most capable and effective teamsters have many identifiable qualities in common. One is that they put a premium on having their animal's harness properly fit. First and foremost is the fit of the collar to the horse's neck, followed by the fit of the hames to the collar, and then on back to the strap configurations and their adjustments.

While it may be true that animals will work in poorly-fit and ill-adjusted harness, most times in those cases the animals will not be able to sustain long hours, day after day, of the discomfort and abrasions. The consequences may include their being less and less willing to pull, and the development of sores and muscle issues. (Some of us believe that prolonged discomfort will cause the animals to think less of their owner.) It takes very little time and effort to make the harness comfortable and the rewards will be long term.

Collars are measured on the inside; top to bottom and side to side, with the top to bottom measurement being primary. Most people are satisfied if the length of the collar is somewhere close to right. If you are interested in maximizing the working efficiency and realizing more of the true art of working horses, great care should be taken to have collars fit perfectly. And not just when the animal is standing around, what counts is when they are leaning into the collar. Positions change, just as with shoes on human feet (you should be standing when your feet are measured and ideally the foot is measured at several points). Collars may take many shapes and should be matched to the shape, thickness and curvature of the horse's neck.

The distance from the withers to the throat, in line with the shoulder, gives us the measurement for the collar. But a perfect fit cannot be acheived without matching the shape of the individual collar to the shape of the individual horse and only then when the collar is employed in pulling. The width of the neck may be accounted for either with the style of collar, i.e. half sweeney versus full sweeney, or by pulling the collar together thru tightening the hame straps, or, in a last resort, getting a collar that is a bit too long, soaking it in water and compressing it to increase the oval.

The first order of business in collar fit is to have the length be as short as possible without pinching off the wind pipe. (I know of no one currently making them but the old "pipe throat" collars were a supreme design for allowing a perfect fit without obstructing breath.)

If the collar is too loose (or too big) several things may occur;
   1. the "draft" (or widest part) of the collar may ride below the point of the shoulder and cause irritation, discomfort and eventual "baulkiness" from the animal.
   2. the collar may rock back and forth as the load is pulled, causing sores to form, especially at the points of the shoulder and top of the neck.
   3. it may cause the animal to, 'counter-intuitively', lift its head and extend its neck in an effort to make the discomfort of the collar fit go away.

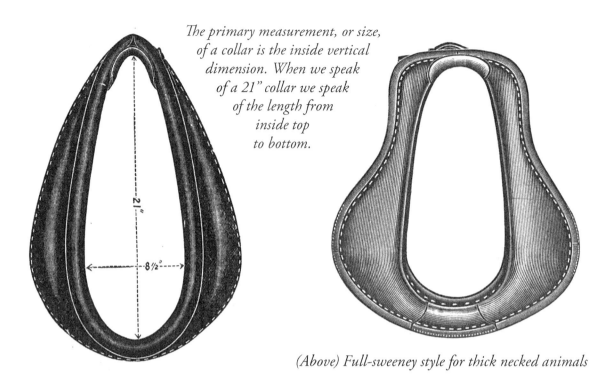

The primary measurement, or size, of a collar is the inside vertical dimension. When we speak of a 21" collar we speak of the length from inside top to bottom.

(Above) Full-sweeney style for thick necked animals

(Below) Half sweeney collars for most draft breeds

If the collar is too tight (as in either too short and/or too narrow);

1. the horse's wind may be restricted, which might cause him to back away from the collar.

2. a pinching of the sides of the neck might also cause the horse to back away, or become a less willing puller.

3. friction will occur causing sore spots to form. Any of these problems will reduce the effectiveness and perhaps availability/suitability of the horse for work.

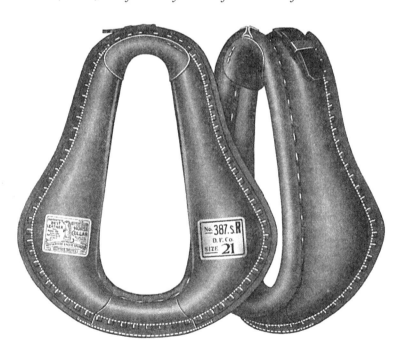

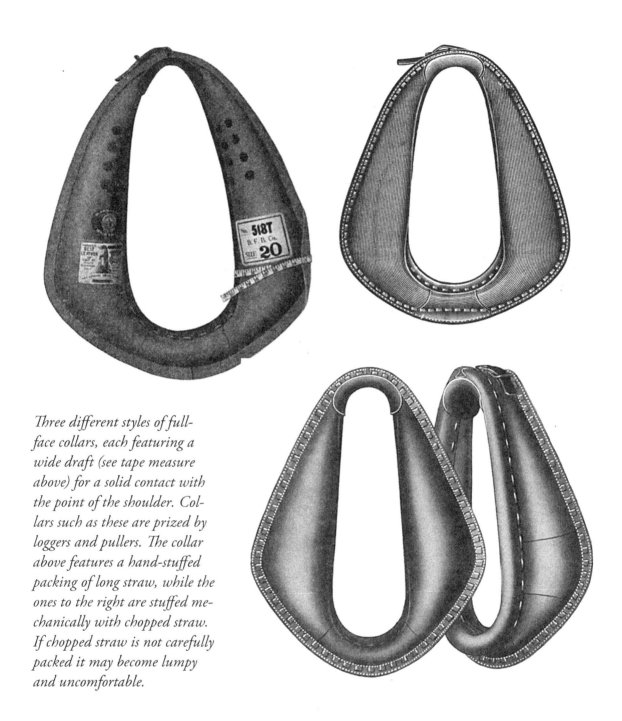

Three different styles of full-face collars, each featuring a wide draft (see tape measure above) for a solid contact with the point of the shoulder. Collars such as these are prized by loggers and pullers. The collar above features a hand-stuffed packing of long straw, while the ones to the right are stuffed mechanically with chopped straw. If chopped straw is not carefully packed it may become lumpy and uncomfortable.

Sore Shoulders: Through years of experience, wherein it was critical to keep horses fit for day-in day-out field work, I tried several tricks and gimmicks to care for their shoulders. Running cold, fresh water over the neck after unharnessing to reduce swelling and clean off salts, or washing down the neck with salt water and a brush to clean shoulder and return salts - those two approaches, popular in their day, always seemed to me to be completely

Top or cap pads are sometimes used to shorten a long collar but that is not advised as it may cause the point of draft to raise up and cause sore shoulders. Another reason for the top cap is to alleviate any discomfort at that point of the neck.

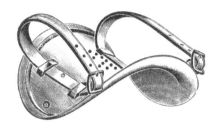

A bridge pad takes all contact off the top of the neck and transfers it back and ahead. These were therapeutic in nature, a hundred years ago, but we don't see them at all nowadays.

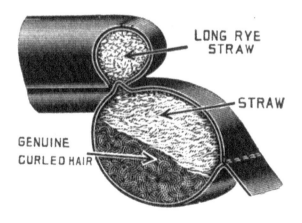

A cutway view of the insides of a collar.

contradictory to one another but good and capable teamsters swore by each. I admit to trying both with what seemed like good results. But still, occasionally, a horse would develop a sore or two on the point of the shoulder or top of the neck, under where the collar would ride, and before you could say "bob's your uncle" the hair was gone and the skin rubbed raw and swollen. Over time, I developed two techniques that saved me days and days of lost work and kept my horses comfortable. Not knowing the scientific jargon to go with the phenomenon, I did pick up on the fact that every single sore was the result of some undue friction and progressed or developed in much the same way. So first order of business was to check every collar surface before and after collars were put on. Running my open hand over them seemed to suffice. If I found something sharp or irregular, I fixed it. Then, with finger tips I would run over the points of the horse's shoulders and the top of the neck feeling for heat, or loose skin, or swelling. If, with slight pressure, the horse backed away or flinched this was something I took note of. If I found a tell-tale spot, I would then go back and carefully check the surface of the collar again. And, if used, I would also check the pad. It could be a piece of gravel or dirt, or a tear in the surface, or a sticker or stiff bit of stitching. It could be something very small or something large. (I am reminded of the old fairy tale of the princess who couldn't sleep on a stack of mattresses for a pea inserted at the lowest level.) If I caught the problem before it altered the horse's skin in any way, and removed the offending problem, the sticker, thistle leaf, foxtail, gravel bit, hair ball, dirt lump, whatever it might be, I could most times avoid it getting to the stage that required taking the animal out of the work string for a time. The second approach was remedial. If after work, harness removed, I found an area on the point of shoulder or top of neck where there was; 1. heat or 2. a wrinkle of skin or 3. a small raised swollen area, I would go get my jar of Ichthymol oinment, a black tar-like paste that worked as a drawing agent. I would cover the area with this stuff and turn the horse out for the evening. In the morning, when I brought the horses in for feed and harness, I would wash that ointment off with a wet rag and, without failure, discover that the heat, wrinkle and/or swelling were gone - every single time! When I caught the problem at the very beginning, removed the offending irritant and used the Icthymol, my horses were always ok to return to work. Amazing stuff. And you should be able to get it at most feed stores and veterinary supply houses.

Important as they might be for you when you feel a problem is coming your way, none of these stop gap measures or miracle solutions replace careful attention to the fit of the collars and to the process necessary to get horses in proper shape and condition before pushing them to long repeated days of hard work.

Conditioning Horses: Way back in the beginning, I often heard teamsters speak of getting *legs* under their horses. Took awhile, but I came to understand they were speaking of the routines and time it took to get horses *fit* for pulling a load all day long, day after day, for long seasons, as the fit of the horses, their conditioning for work, prepared them for each and

A quiet, willing, single horse in open-bridle pulling a walking cultivator in Germany. One way to measure the certainty and training of this animal is to notice that the rope line back to the teamster is completely slack and yet the animal walks confidantly and straight ahead through the plant row. Photo by William Castle.

every pressure, from collar fit to hoof traction, from neck and head comfort, to the balance of their frame moving against a load. Any and every discomfort is a potential distraction and therefore a safety hazard to the teamster and a point against ultimate training. The steps

A handsome team of Belgians does the cultivation work for Steve Hagen at Oregon's Old School Vineyards, photo by Harley Hagen.

taken to build up muscle tone, lungs, and joint health are very much akin to what you might imagine a human athlete does. Excercises employed in a routine or pattern, slowly advanced and advancing, these allow the horses to ease into peak condition. There are many ways to accomplish this. I liked to use my horses in the winter for feeding, manure spreading and light work, keeping track of how long they spent actually exerting themselves. I found, for most individual horses, from four to five days of one hour per morning and one hour per afternoon allowed me to advance to three hours in one stint each day for a week (maybe field dragging or light discing), ratcheting up to two three hour shifts, morning and afternoon (doing about anything - from plowing, to discing, to harrowing) the third week, to finally at the fourth week eight hours a day with an hour and a half midday break. When I got to this point my horses had "legs under them" and I knew that in a pinch, say a storm coming and I had to finish seeding a field, they could be called upon to do some overtime. Easing horses into full-time work in this way took time and trouble but the results were fantastic. And I seldom had any problems with sore shoulders. Thinking and writing about this stuff I find myself wanting to say that it starts "here" or "there" but I know deep down that everything is important. While conditioning is a real useful secret, so is harness fit.

Sweat pads fasten to the inside of the leather collar, and, if properly constructed, provide a cushion for the horse's shoulders. They are made of either deer-hair ticking or slabs of felt. (There are modern synthetic pads available which soak up little or no sweat.) Pads are sometimes employed to make a big collar fit better. Their size is measured top to bottom, and because they follow the curvature of the collar, they are typically 2 inches longer than the size of the collar. For example; if the collar is 24" in size, you will need a 26" sweat pad; that collar then will effectively fit a 22" neck.

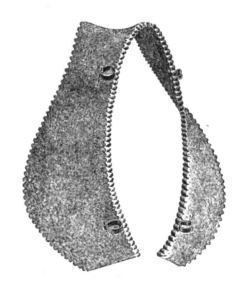

William Castle of Shropshire England works his Percheron mare on a one-horse McCormick mower. Note that William uses a collar pad. Aside from a pad allowing that a too-big collar may be made to fit properly, there are many opinions regarding whether or not the collar pad adds to the animal's general comfort. For extended periods of work, this author is convinced that the clean, properly-fit collar pad lends a great deal of comfort to the animal. Also note that William's harness differs substantially from the North American varieties, with a heavy cart-style back pad and rope driving lines. Look to page 47 for more images of harness variations from Europe.

Notice separately; in this photo, you can clearly see that a chain passes over the back pad or saddle and fastens to the shafts to hold them up on each side. Up from the belly band, notice a shaft loop has been formed to hold the shaft down.

Ray Drongesen driving his mare Ruby in the shafts on a 'stud cart' (so named because it was common for horsemen to travel from farm to farm with their stallion in cart shafts. This proved to be an excellent advertisement for breeding services.) This photo was taken in Junction City, Oregon, in 1975. The children in the cart are Ian and Juliet Miller. Several things to note in this picture: The harness fits the mare perfectly - see how the collar seems to be a part of the neck and shoulder with the length perfect at the throat. Also notice the hip straps and position of the brichen: there is no slop apparent anywhere. Ruby looks like she is able to move in any way without feeling the slightest restriction or discomfort from the harness. The shafts are at the exact right angle and held there by a shaft loop made of a hame strap through upper and lower billet strap hardware, from belly band and neck pad. Ray, Ruby and the entire setup tell the story of a perfect match. Photo by Nancy Roberts

Thinking of that phrase "comfortable in his own skin." I like to ask, with each animal, is he or she comfortable in the harness, or perhaps more accurately, does the harness appear to be a perfect extension of, and natural for, the animal. That starts with the collar and then moves back to the skeletal structure of the harness.

Far and away the most prevalent harness design employed in North America is the western basket brichen (with either a two or three strap brichen). Because of the imprecise and variable adjustments possible with the quarter-strap arrangments, this style of harness is more forgiveable of poor adjustments.

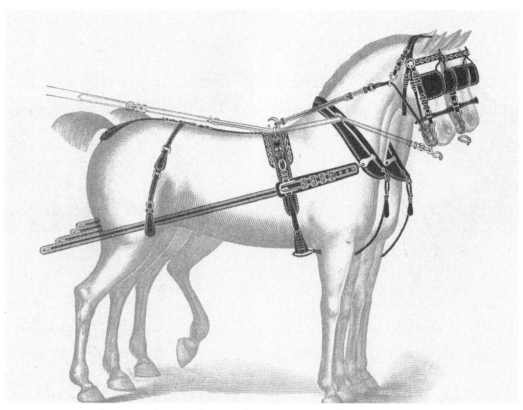

A fancy coaching style team harness not suitable for farming, logging or heavy work; it is too light in all points.

Draft pony breast-collar style buggy or cart harness, designed to be used with shafts.

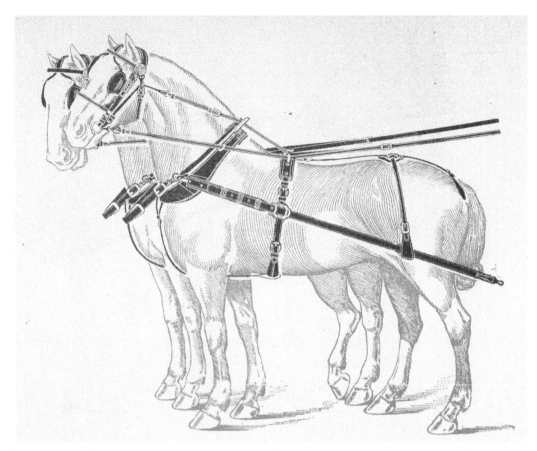

What is frequently referred to as a plow harness, designed without a brichen assembly. Some modern harnesses come equipped with a detachable brichen to remove unnecessary parts when utilizing a walking plow or when logging. Often, in big field hitches where only the wheel team (those animals closest to the implement and hitched to the tongue) provide direct braking and backing, efficient farmers will use these lighter weight "plow harnesses" for the remaining animals in swing and lead positions.

The ***New England D Ring and Boston Side-backer*** styles of harness require far greater precision of adjustment for them to work properly.

A good rule of thumb would have you pay attention to the angle and line of draft (or any restraint or 'broken line'); the line of draft should be unbroken or straight from one end to the other. (See page 143.) If a brichen is too low, or the hip straps too short, when backing or braking there is downward force on the hips causing discomfort. The same can be true of traces running through lazy or mud straps that are too short.

The brichen encircling the lower rump level or with a slight forward incline is the most common design and adjustment. In some western parts of the continent, teamsters preferred

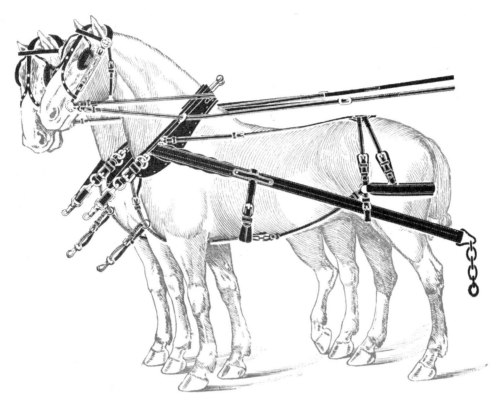

Two styles of express or light working harness, the one above without back pad.

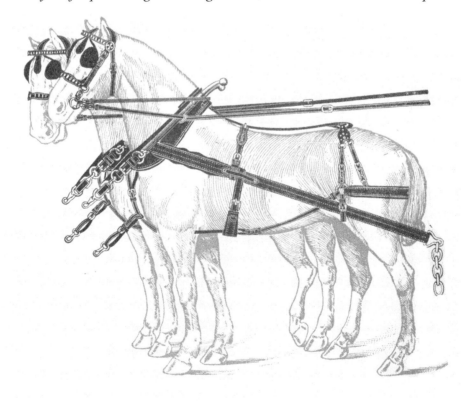

what is referred to as a Yankee brichen which had the heavy strap run at a severe angle, crossing over above the tail seat. This design allowed for less weight and material but does have specialized application restrictions or discomforts.

There are different opinions when it comes to how belly bands are to be adjusted. Some like them snug against the belly, while others (this author included) prefer them to hang loose (two inches from the belly). Here is an area where you should allow your experience and observations to guide your choice. Another element in that mix is how, with the western basket brichen harness, you position the pole strap (which connects in a Y to the two quarter straps). If you allow this to hang free below the belly band, there is less opportunity for a broken line of restraint when backing. If you thread the pole strap between the belly and the belly band, and the belly band is snug, you may transfer unnecessary pressure to the back pad while backing. (In some cases the pole strap features a loop through which folks thread the belly band adding a slightly different dynamic.)

D Ring Harness, note the two part tug and sidebacker straps jointed at ring.

Jason Rutledge with side-backer harness.

A side-backer harness being used in Germany, on a ground-drive forecart and hay tedder. Photo by William Castle.

In New England draft circles, the name of Les Barden rings clear. He was a stalwart champion of good horses well-harnessed. Les believed in the "D" Ring harness above all others and preached its virtues far and wide.

The "D" Ring Harness *by Carl Russell*

The harness that Les Barden used is the D-ring harness, or New England harness, or New England D-ring harness. There is a similarity to the Sidebacker harness insofar as they both use a pony yoke style pole yoke, with side straps carrying the hold-back load from britchen to yoke. The main difference is that the Sidebacker had side straps that went all the way from the britchen ring to the yoke, whereas the D-ring has rear and front side straps that attach centrally at the D-ring. The sidebacker was also called the Boston-backer, or Boston Sidebacker, so the two were clearly New England-based harness systems, and there is some suggestion that the D-ring may have resulted from a modification of the Sidebacker. Here is one of the most famous photos of Les working his greys in the D-ring harness, Barden Log/forecart, and his plug-style yoke at a logging gathering at Earthwise Farm and Forest in 2005. Comfort, Safety, and Dignity were primary principles of Les' working philosophy.

- Photo by Lisa E. McCrory.

Les Barden was a remarkable individual. Years ago, when he was still alive and in his eighties, I spent a day and a night with Les on his New Hampshire tree farm. He made a powerful impression on me. His place was spartan and correct by his unique and adamant design. Everything in its place, nothing superfluous. (Unless you made the mistake of considering his pigeons as superflous.) His firewood was precisely and exactly the right length and thickness, and it was stacked with excruciating precision. His sawmill, wood lot and log deck were organized and orderly. His team of greys, standing comfortable in their tie stalls. exuded good health and fitness. In the presense of this good and great man it was impossible not to feel the clarity and effectiveness of his self-imposed discipline and clean living. LRM

COMFORT, SAFETY, AND DIGNITY

by Les Barden of Farmington, New Hampshire
originally appeared in Small Farmer's Journal Volume 30- Number 1

This picture (page 49) was sent to me recently and it speaks the proverbial thousand words. Here under close examination will be found, for the horse and driver, the features of comfort, safety, and dignity.

The heads are liberated from unsightly and, too often, cruel halters, which interfere with bridle fit and function. From straight bits, which hold squarely against the jaw, rather than pinch the sides, as do jointed bits, the reins run through drop rings, which are four inches lower than the fixed hame ring. The object is to hold the bit against the jaw, not cause it to pull up into the corners of the mouth. High-headed horses do not need these drop rings. By all means, check reins should be run if needed.

Since the collars fit satisfactorily, pads are not used. Orchard hames are of no consequence to the horse. However, in logging the driver is spared the shower of snow from boughs and the whip of branches that often occur with hame knobs.

There are no lug straps from the hames to the twin yoke. All tongue and yoke weight is on the back pad when the D-Ring harness is adjusted properly. The thin, long lazy-straps, shown here, never come into play except when hitching and unhitching. They then keep the side straps from hanging only from the D-Ring. The short, forward tug leaves the hame bolt at a 90-degree angle. It is locked in position by the D-Ring, which is secured by the correctly adjusted market straps and girth. The collar will not ride up or down regardless of the angle of pull on the rear tug.

The breeching is located high on the rump just below the pin bones. It is snug, but does not indent the muscles unless the horse is holding or backing.

Horse standards have been applied by the driver in designing the log cart/forecart. The driver stands safely on a walk-through, gravel-screen deck, which is sixteen inches from the ground. A steel dashboard, dash rail, and toolbox are conveniently placed before him and a high full cushioned seat and a steel bumper-plate are behind him. A unique feature is the one-handed quick hitch and unhitch arrangement of a slip and grab hook. Weighing about 350 pounds and being only 57 inches wide it is easily loaded up ramps into a pickup by one man.

Attention is called to the pole end, which does not protrude beyond the twin yoke. Attached to the yoke is a clevis from which extends a 5/8 inch x 5 inch pin that slips into the end of the tongue. The tightness of the D-Ring harness within the rigging (not upon the horse) keeps the plug securely in the end of the pole. The driver designed this for, and intends it only for, use when the D-Ring harness is properly adjusted.

To avoid negating the correct adjustment of the brace reins, the driver keep the long reins between the horses at all times. Drop rings from the breeching center suspend the reins so that when they are not in hand they do not sag. On equipment with long poles such as sulky plows and mowers, a 20-inch connection strap can be used between the drop rings to prevent a rein from riding over a horse's back while turning sharply.

Not visible are brass screws securing the flap ends of market, side, bit, and bridle straps. This offers a neat appearance. The absence of superfluous harness parts and decorations along with an exacting fit of the harness displays the artfulness of a well-presented horse, which is turn, dignifies the presenter.

And comfort means fit in all respects: fit as in muscle tone and readiness, fit as in right for the work, but in this case I primarily speak of the harness and how it is adjusted to "fit" the horse. A loose, saggy, dangling harness advertises what to expect from horses and teamster. A collar that is too long and wide for efficient work says "this horse is not going to work efficiently." In time this horse will become less and less comfortable and finally damaged.

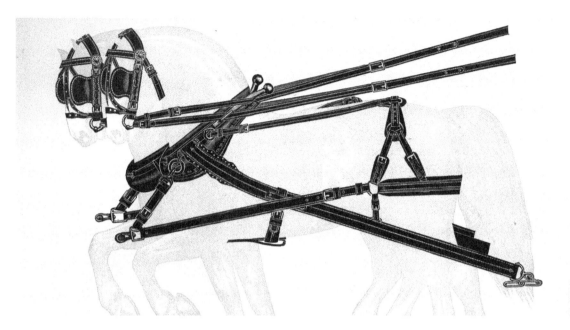

Two versions of a heavy-duty side-backer, or New England, harness.

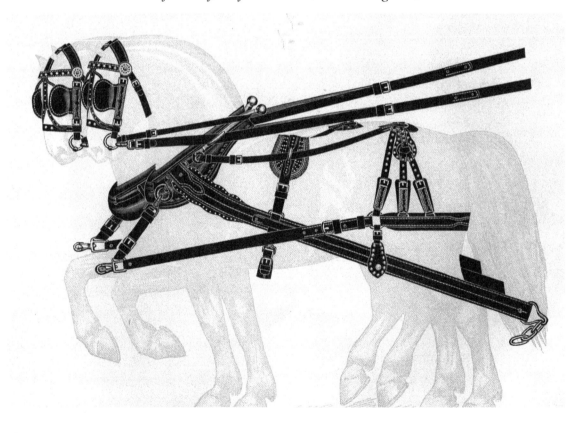

Note in the upper picture the 'market-tug' assembly, utilizing a triangulated back strap running from hame to tug billet and back up to crupper.

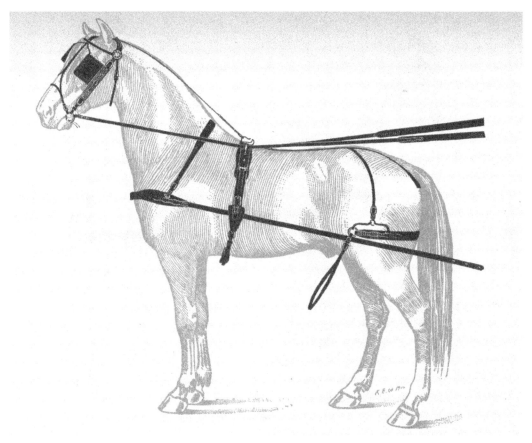

Breast-collar style buggy or light wagon harness.

HARNESS VARIATIONS FROM FAR AND WIDE

For three thousand years or more humans have 'asked' animals to pull stuff for them. In almost every case, some sort of harness was employed to transfer forward motion to a pulling action. Aside from the distinct peculiarities applied to oxen (sometimes tying heads together), camels (pulling from the hump), buffalo, and ostriches, most of the harness mechanics, as seen with horses and mules, follow basic principles. Either the neck, chest, or shoulder provides the push with breast or neck collar, and a web of straps hold it all together. In North America, we have seen less variety in the basic design than elsewhere in the world.

The ancients figured out eons ago that willing mammals might be put to the job of dragging a sharpened stick through the land, providing basic tillage ahead of planting seed.

All around the arid edges of the verdant middle eastern cradle of civilization people called upon camels to scratch the dry, hard ground.

In the beginning it must have been a challenge...

but with the passage of time, and the handing down of skills, the Arab peoples became more and more adept at the process of farming with draft animalas. Note in these two pictures the extreme angle of draft and that the point of draft begins with a pad on the front of the camel's hump.

These beautiful photos by William Castle portray the ornamented design of European harness. Note the massive one-piece collars on the logging team.

Mattias Brüning's Rheinisch –Deutsch gelding waits patiently while Klaus Strüber explains the features of Albano Moscardo's cultivator, seen here holding the handles. Photo by William Castle.

This French working team leans with certainty into the pull. Their highly unusual harness begins with collars which include, in their construction, the hame structure and a padded cap with guide holes for the driving lines. The tugs are sheathed chains. There are no belly bands, back pads, or brichens. A cruper around the tail secures the back strap from which hang two trace carriers. Photo by Jean Christophe Grossettete.

This lovely French Comtois gelding carries a new style of combination collar, and a mechanical hackamore instead of a bit in the mouth. Photo by Jean Christophe Grossettete.

William Castle took this photo at Pferde Starke, Germany, of a highly unusual harness design which features triangulated tugs that meet beneath the belly and pass through the hind legs to the implement being drawn.

Two European show-style harness riggings, above with breast collar, below with combination collars. In both cases the attractiveness does not diguise the general discomfort of both designs. Long days of work would result in sore shoulders. Photos by William Castle.

A most unusual Scandanavian harness design featuring what appears to be an upside down collar hung round the neck to carry the end of the tonuge plus a heavily padded breast collar. Photo by William Castle

Jean Christophe and apprentice driving a Comtois gelding in a modern combination collar.

Frenchman Giles Marty with his Ardennes stallion drawing attentive crowds in the logging competition; note where the lines pass through the tops of the hames. Photo by William Castle.

This scene from Detmold, Germany, also by William Castle, shows a very unusual collar/hame setup.

When the imagination meets the tractive possibilites of horsepower.... ?

The original and so far only, two-horse dump truck, steered by Merle Ligocki of Wyoming. Nick and Buck providing the motive power. Photo by Dale Ackels.

HARNESS PARTS

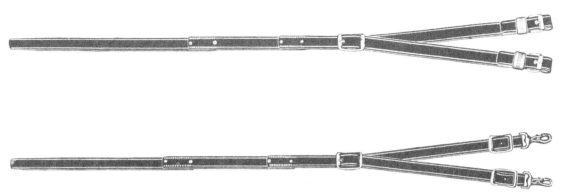

Team lines (two to a pair or set) customarily offered in widths of 7/8", 1", 1 1/8", or 1 1/4" with lengths of 18', 20' 26' or 32' (most common length 20'). Above; set to buckle into bits (safest). Below; set to snap into bits. Note cross checks affixed by full buckle (avoid lines with conway buckles for cross checks).

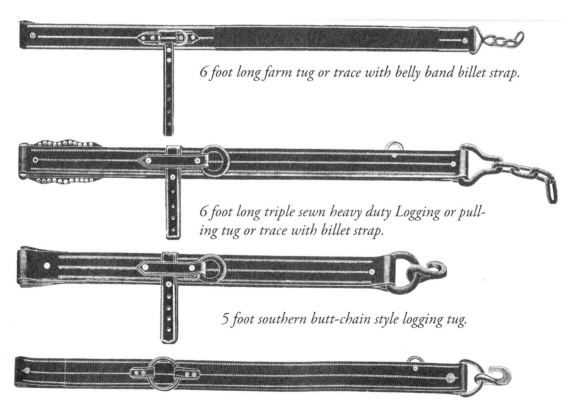

6 foot long farm tug or trace with belly band billet strap.

6 foot long triple sewn heavy duty Logging or pulling tug or trace with billet strap.

5 foot southern butt-chain style logging tug.

6 foot butt-chain billet ring tug

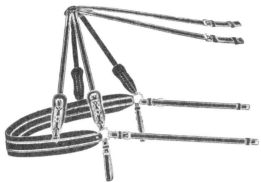

Brichen or breeching assembly including back straps, quarter straps and lazy straps. The back straps pass through back pad and attach to second rings on hames. the quarter straps run under the belly of the horse and attach to the pole strap, which in turn runs forward to the end of the neckyoke. The tugs or traces pass through the lazy straps enroute to the single tree.

Spreaders are employed when we want our horses to work a little further apart. They hang from the top ring, or hame ball, on the inside. The cross checks pass through the end ring, elongating the triangles formed by the lines, and effectively moving the horses futher apart.

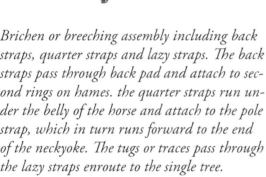

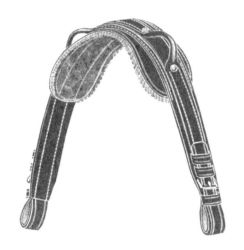

Center rings, or "hearts", are sometimes used and will offer some "cleanup" of the dangle of the cross checking lines. They tend to tidy-up the dangle and keep things organized.

HARNESSING with a Western-style basket-brichen farm harness.

(These photos were taken of Ray Drongesen harnessing King way back in 1976, which goes to show that the process hasn't changed.) 1. Either tie up your horse or have someone hold it by lead rope. If collaring is new to your horse, or you are new to collaring, allow the horse to look at and sniff the collar first. Then, open the collar at the top. Brace the opposing side with your left hand, your right hand on the near side of the collar, and slide the collar up the horse's neck as pictured. (Note: Many horses may learn to have the closed collar slipped over their face, head and ears without opening. If you are unfamiliar with the horse or the animal is new to harnessing postpone this method until confidence is reached by all parties.)

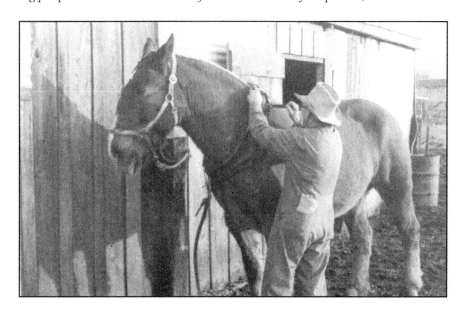

2. Fasten the collar at the top, making sure no dirt or hair bunches or debris is between the horse and the leather face where it might cause an abrasion.

3. In this photo, Ray is checking the throat of the collar to make sure it isn't too tight. A collar must fit perfectly if you are to avoid collar sores. If the collar is too tight at the throat, as the horse pulls it will cut off its own wind.

4. Ray's right hand and arm have been passed under the brichen and back pad between the two tugs, and he is holding the right side hame at the mid-point. His left hand is holding the left hame at the mid-point. Harness held in this way allows you to separate it as you lift to place on the horse's back.

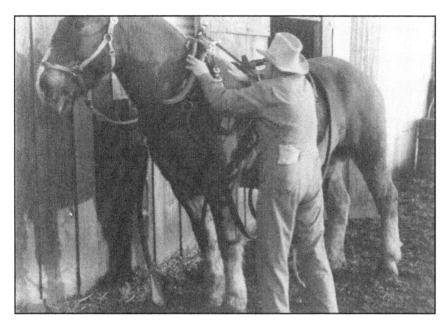

5. Ray has passed the harness over the horse, hame in the vicinity of the collar groove, and is backing his right hand out from under. He has placed the bulk of the harness on King's mid-section, balancing it so that it does not fall off the right or left side. The brichen assembly is NOT pulled back over and under the tail at this time.

6. Ray pulls the harness back over King's torso, with hames seated loosely in the collar groove. With a nervous or inexperienced horse it is important to prevent the harness from falling back and off the horse, as it may cause the animal to jump or kick out. If the hames aren't seated in the collar groove when the harness is pulled back, there will be little or nothing to restrict the harness from falling back.

7. The hame strap, at the bottom of the hames, is tightened firmly and buckled, making sure that both hames are seated snugly into the goove on the collar. If this buckling is not done tight, the hame strap can pop out of the groove when the horse is pulling or straining against a load. This then would shut off the windpipe. Pull back hard on the strap...

8. ...and buckle tightly.

9. Next buckle the belly band, allowing at least a hand's width between the band and the belly, with room for the pole strap to pass through. The breast strap is then snapped on the horse's left hame at the bottom ring, forming a V. The pole strap is either hung by a string from the collar throatt, or fastened by a combination snap and slide, to the middle of the breast strap.

10. One from each side, the quarter straps are brought forward under the belly and snapped to the end of the pole strap. The quarter straps originate at the ends of the brichen.

11. Here Ray checks to make sure that the quarter straps have just the right amount of tension in them. If they are too loose a horse can step over them when swatting at flies. If they are too tight, they put an uncomfortable pressure on the belly when the horse is asked to brake, back, or hold a load.

12. King is harnessed except for his bridle.

13. This photo illustrates the perfect position for the brichen strap allowing that the quarter straps flow in an unbroken line from the end of the brichen on each side in a loose fitting curve, following the belly of the horse forward to the pole strap.

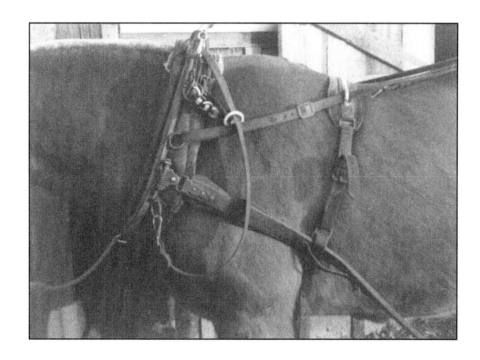

14. This photo shows the perfect fit of the working collar with the harness in place.

15. With the throat latch undone, the bridle is pulled up the face and spread to allow the bit to approach the mouth. Some horses willingly open their mouths to receive the bit, others will require a slight coaxing by placing thumb and finger in opposite corners of the mouth to help them open up.

16. While raising the top of the bridle, the bit is slid carefully into the mouth, avoiding any quick jerking that might chip a tooth or give the horse cause, through discomfort, to worry about this process the next time.

17. Ray gently passes the top of the bridle over King's ears. He knows that it is important not to allow this procedure to result in a horse being bridle shy. If your horse has a long mane or forelock, make sure the hair isn't pulled over the ear in a fashion that might cause the horse to want to rub it off.

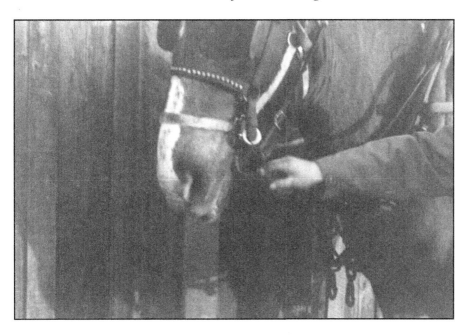

18. A quick check to see that the bit is comfortable in the gelding's mouth. Note: in this instance Ray has elected to leave the halter on under the bridle. It is important that the halter fit well and not dangle and tangle. The halter is always a better thing to tie a horse to rather than a bridle or bit. Some teamsters, however, feel that leaving a halter on is wrong and a show of laziness. To each their own.

19. Here he is attaching the check rein. Some harnesses, such as this one, have a strap and snap coming forward from the brichen spider to receive the check rein. Others prefer to loop the check rein over the ends of the hames.

20. King, all harnessed and bridled, and wants to know what all the extra fuss is about.

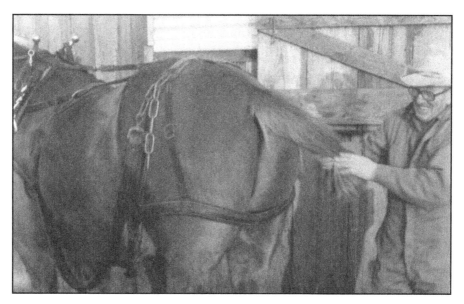

Ray never tired of a good horse well harnessed.

Photo taken by the author at the National Plowing Championships in Dalton, Ohio.

It's all about the angles. Here the Adam's family Percherons demonstrate the important parallel lines and what they might tell us. Note the angle of the collar on the shoulder and its relationship to the forelegs as they move ahead. See the angle of the faces and their alignment with the forward cannon bones? When these parallels are absent we see discomfort and imbalance. These plowing horses are exerting themselves, but with perfect balance and comfort.

The fit of these bridles, as well as the head set, tell us that these mules are comfortable. Note that the throat latch is snug as it should be. Photo by Kristi Gilman-Miller.

EVENERS

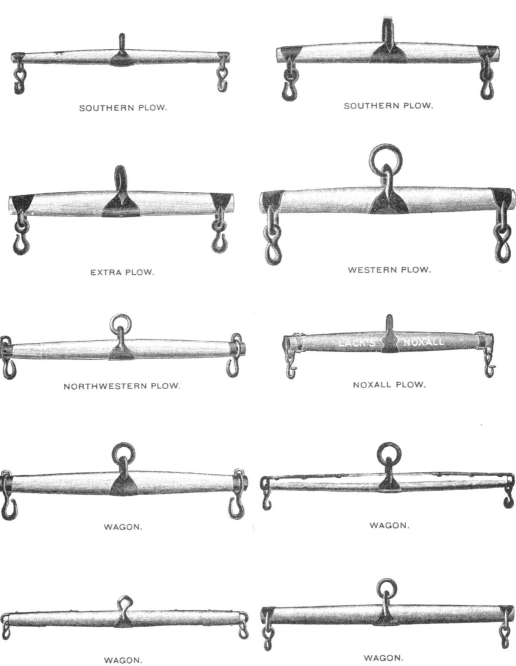

These next seven pages are reprinted from a 1910 hardware catalog.

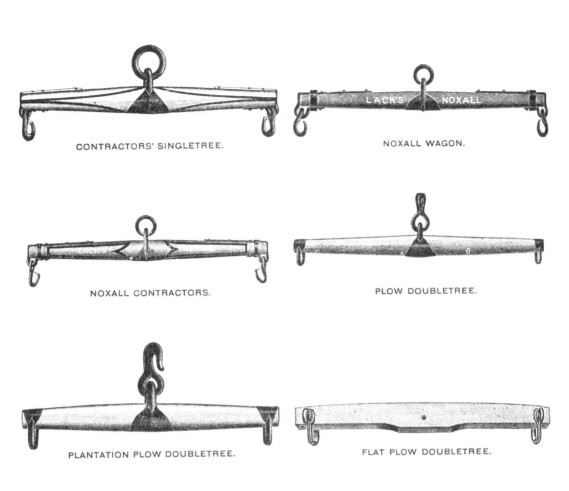
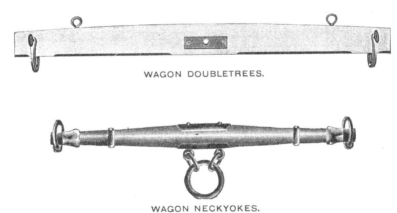

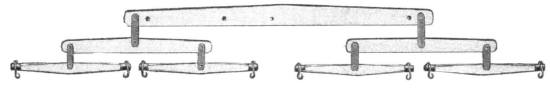

Cut showing four-horse.

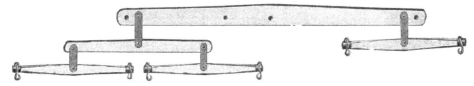

Cut showing change to a three-horse.

PHICO COMBINED TWO, THREE AND FOUR-HORSE EQUALIZER.

SOUTHWESTERN TRIPLETREES.

WESTERN TRIPLETREES.

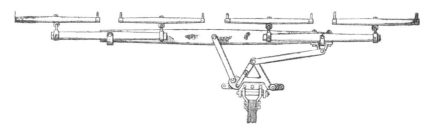

HEIDER'S FOUR-HORSE PLOW EVENER.

TRIPLETREES AND EQUALIZERS.

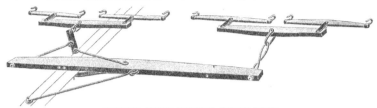

OWENS FOUR-HORSE EQUALIZER.

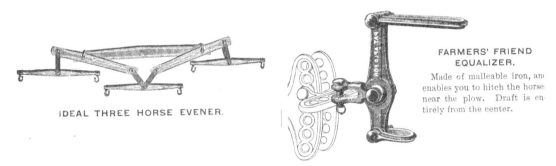

IDEAL THREE HORSE EVENER.

FARMERS' FRIEND EQUALIZER.

Made of malleable iron, and enables you to hitch the horse near the plow. Draft is entirely from the center.

TUBULAR STEEL DOUBLE AND SINGLETREES.
EQUIPPED WITH DROP FORGED TRIMMINGS.

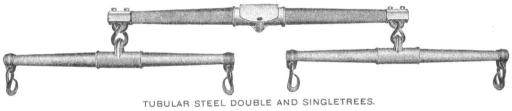

TUBULAR STEEL DOUBLE AND SINGLETREES.

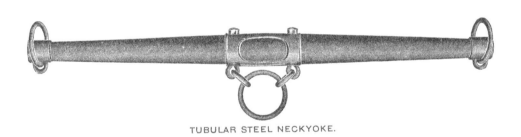

TUBULAR STEEL NECKYOKE.

TUBULAR STEEL SINGLETREES.

TUBULAR STEEL DOUBLE AND SINGLETREES.
EQUIPPED WITH DROP FORGED TRIMMINGS.

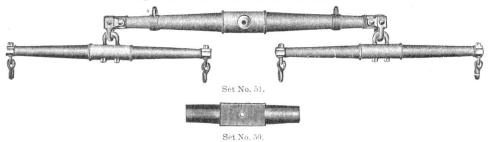

Set No. 51.

Set No. 50.

TUBULAR STEEL WAGON AND TRUCK DOUBLETREES.
FOR PLATFORM WAGONS AND TRUCKS.

TUBULAR STEEL DOUBLETREE.

New Pattern No. 70, with Drop Hooks.   New Pattern No. 71, with Cockeye Hooks.

TUBULAR STEEL SINGLETREES.

TUBULAR STEEL SINGLETREES, WAGON OR TRUCK.   TUBULAR SINGLETREES.

TUBULAR STEEL DOUBLE AND SINGLETREES.
NEW PATTERN 102. FOR FARM WAGONS AND FREIGHTERS.

TUBULAR STEEL NECKYOKES.
NEW PATTERN 103. FOR FARM WAGONS.

TUBULAR STEEL SINGLETREES.
NEW PATTERN 104. FOR FARM AND FREIGHT WAGONS.

TUBULAR STEEL DOUBLETREES AND SINGLETREES.
NEW PATTERN 105. FOR PLOWS.

TUBULAR STEEL SINGLETREES.

TUBULAR STEEL SINGLETREES.

TRIMMED NECK YOKES

PEARSON'S.
Swivel center, and clasp, acorn ends, 42 inches long.

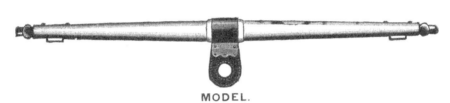

MODEL.
Steel band, leather drop, acorn ends, malleable loops, 42 inches long.

SHUMAN; SWIVEL.

SOVEREIGN; SWIVEL.

ROCKFORD; SWIVEL.

AUTOMATIC.

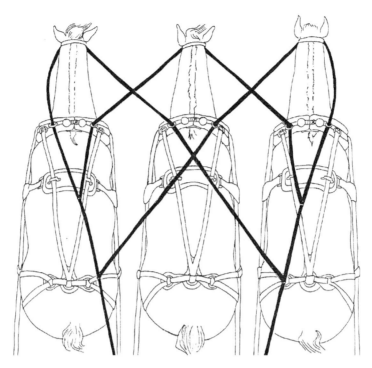

Setting up lines for three abreast, you can use a special three abreast line configuration with a second, longer check on each line, thereby having a direct line to the end of every bit. Or, as seen on the next page, you can employ stub checks that fasten dead to the opposing horse's hame ring. The same thing is true in the four abreast below. The author prefers this setup and feels that it provides greater precision.

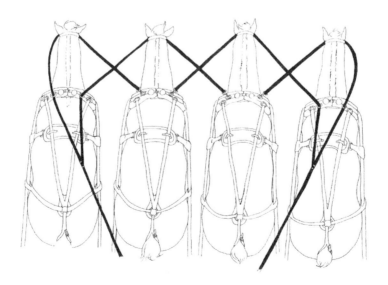

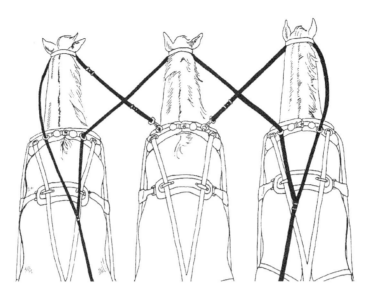

I have learned that driving three or four breast is often easier that driving a single or team. I have a theory to explain why that may be true. The single horse has no other equine to 'lean' on. With a team, there is always a chance that one will *'infect'* the other with nervousness or fear. With three or four abreast, there seems to be a sense of *'security in numbers.'* Set the lines properly and give it a try.

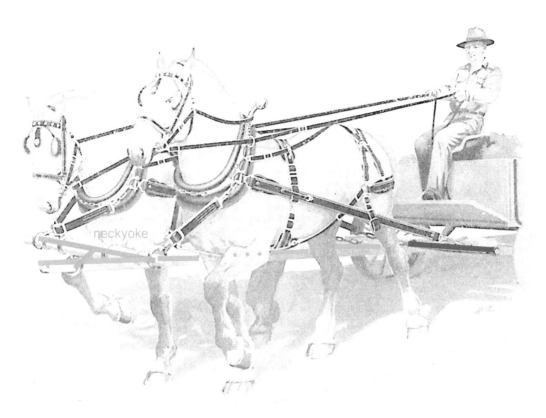

ALWAYS hitch the neckyoke (blue) to the front of the tongue or pole (green) before the singletrees (red).

Here I am hitching Bob and Granite to a mower. Difficult to see, but I do hold the lines in one hand as I hook black Granite's inside tug. I hook the inside of each singletree first so that, if the team should step ahead, before I get done, the pull will be even (with less chance of slamming and tangling of the unhooked singletree). Certainly not necessary with this excellent, well-behaved team, but a good habit to practise.

With hitching and unhitching, I have found that it is extremely useful to always follow the same routine.

AND another important observation: I do not let my horses move, when hitched or unhitched, until I am ready for them to move. In fact, I like to make them stand, especially when unhitched, until they relax. The horse acting on anticipation will often result in a problem. So I practice taking their anticipation away from them.

Jim Butcher with the master's touch (Dayton, Ohio). Photo by the author.

Chapter Two

PROGRESS

Where Were You When She Learned That?

Queenie, a two thousand pound Belgian mare, was good to handle, harness and drive. But when it came to her hoof care it was always a challenge. This was aggravated by the fact that she had developed a proud flesh growth on her right hind coronet band following a scrape. It meant I had to do regular doctorings and injections directly into tumor, which hurt or irritated her. Getting her to give a hind leg for any care was a risky challenge, until a curious set of circumstances offered a way forward.

One morning I went out to the barnlot with my trimming tools determined to work on Goldie and Queenie's hooves. To my surprise I found Queenie had tangled herself up in a rope and was laying down. It was apparent she had been down for some time. My first thought was that she was dead. When I saw she wasn't I carefully untangled her legs, put a halter on her, pulled her head up and, after a few minutes coaxed her back up on her feet. She was a little dizzy and disoriented but seemed fine otherwize. I don't know what possessed me at that point, perhaps I was still young and stupid, but I figured "what the heck?" With her standing untied and sleepy-like, I lifted her right hind leg and set to work trimming the hoof. She never moved an inch. I did the left hind leg next. Same thing. Did the front feet, her still standing with a halter on and no lead rope. Scratching my head over what had happened, I repeated my moves and went all the way around, lifting each leg. Then I went around the other direction doing it again. Perhaps I was being a little silly because I was enjoying this. After all, she had been such a handful with her feet that this felt like a once in a lifetime opportunity. I smiled and pulled the halter off and left her standing there.

I worked on Goldie's feet noticing that Queenie never moved a muscle. When I finished I returned to Queenie and stroked her neck, starting to get worried about her. She seemed okay, just dopey. I went back to her hind feet and lifted them again. No problems. Went around her both ways lifting each leg. Finally I put the halter back on, snapped on the lead rope and lead her around. All was fine, I turned her loose. After that experience she never gave me a lick of trouble with her feet. But if a veterinarian came around no one could get those feet off the ground.

This good material is reprinted from *A Textbook of Horseshoeing for Horseshoers and Veterinarians* by A. Lungwitz written in 1884. It is good information and most definitely worth a read.

Raising and Holding the Feet of the Horse to be Shod

This can always be done without much trouble if the horse has been accustomed to it from early childhood. Certain rules governing the manner of taking hold of the feet, and of afterwards manipulating them, are of value.

A shoer should never grasp a foot suddenly, or with both hands. The horse should first be prepared for this act. First see that the horse stands in such a position that he can bear this weight comfortably upon three legs. This is well worth noticing, and if the horse does not voluntarily assume such an easy position, move him gently until his feet are well under his body.

If the shoer, for example, wishes to raise the left fore foot for inspection, he stands on the left side facing the animal, speaks quietly to him, places the palm of the right flat upon the animal's shoulder, and, at the same time, with the left hand strokes the limb downward to the cannon and seizes the cannon from in front. With the right hand he now gently presses the horse towards the opposite side, and the foot becoming loose as the weight is shifted upon the other leg, he lifts it from the ground. The right hand now grasps the pastern from the inside followed by the left hand upon the inside and the right hand on the outside; then, turning partly to the right, the holder supports the horse's leg upon his left leg, in which position he should always stand as quietly and firmly as possible. If, now, the shoer desires to have both hands free to work

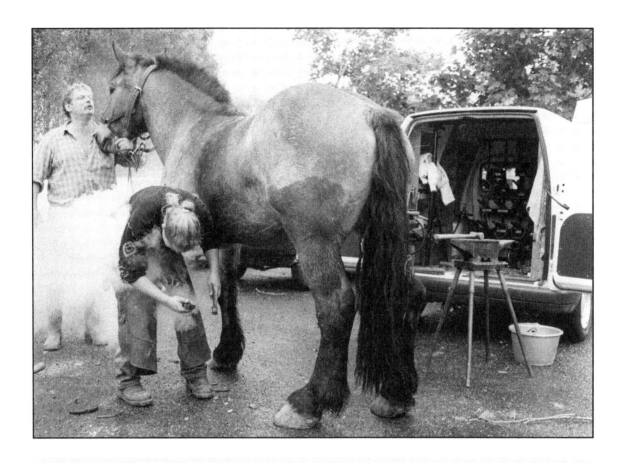

upon the hoof, he grasps the toe with the left hand in such a manner that the toe rests firmly in the palm while the four fingers are closely applied to the wall of the toe, takes a half step toward the rear, passes the hoof behind his left knee into his right hand which has been passed backward between his knees to receive it, and drawing the hoof forward outward and upward supports it firmly on his two knees, - the legs just above the knees being applied tightly against the pastern. The forefoot should not be raised higher than the knee (carpus), nor the hind foot higher than the hock, nor either foot be drawn too far backward. The correct standing position of the shoer or floorman while holding a front foot is shown in *illustration*. Shortness of stature (5'-5'6") is desirable in a floorman.

In lifting the left hind foot the animal should be gently stroked back as far as the angle of the hip, against which the left hand is placed for support, while the right hand strokes the limb down to the middle of the cannon, which it grasps from behind. While the left hand presses the animal's weight over towards the right side, the right hand loosens the foot and carries it forward and outward from the body so that the limb is bent at the hock. The holder then turns his body towards the left, brings his left leg against the anterior surface of the fetlock-joint, and carries the foot backward, at which time his left arm passes over the horse's croup and above and to the inner side of the hock. Finally, both hands encompass the long pastern.

If the right feet are to be raised, the process is simply reversed.

In raising the feet no unnecessary pain should be inflicted by pinching, squeezing, or lifting a limb too high. The wise shoer avoids all unnecessary clamor and disturbance; quiet, rapid, painless methods avail much more. In dealing with young horses the feet should not be kept lifted too long; let them down from time to time. In old and stiff horses the feet should not be lifted to high, especially in the beginning of the shoeing.

Vicious horses must often be severely handled. Watch the play of the ears and eyes continually, and immediately punish every exhibition of temper either by jerking the halter or bridle vigorously, or by loud commands. If this does not avail, then if soft ground is at hand make the horse back as rapidly as possible for some time over this soft surface; it is very disagreeable and tiresome to him. To raise a hind foot we may knot a strong, broad, soft, plaited band (side line) into the tail, loop it about the fetlock of the hind foot, and hold the end. This often renders valuable service. The holder seizes the band close to the fetlock, draws the foot forward under the body, and then holds it as above described. The use of such a band compels the horse to carry a part of his own weight, and at the same time hinders him from kicking. Before attempting to place this rope or band about the fetlock, the front foot on the same side should be raised.

The various sorts of twitches are objectionable, and their use should not be allowed unless some painful hoof operation is to be done. The application of the tourniquet, or "Spanish windlass," to the hind leg is equally objectionable.

Those horses which resist our attempts to shoe them we do not immediately cast or place in the stocks, but first have a quiet, trustworthy man hold them by the bridle-reins and attempt by gentle words and soft caresses to win their attention and confidence.

Ticklish horses must be taken hold of boldly, for light touches of the hand are to such animals much more unpleasant than energetic, rough handling. Many ticklish horses allow their feet to be raised when they are grasped suddenly without any preparatory movements.

FIG. 43.

Right forefoot viewed from the side: *A*, lower end of the cannon; *B*, fetlock-joint; *C*, long pastern; *D*, coronet; *E*, hoof; *F*, heel; *F'*, inner heel; *G*, foot-lock covering the ergot.

Star Horse Shoeing Rack.

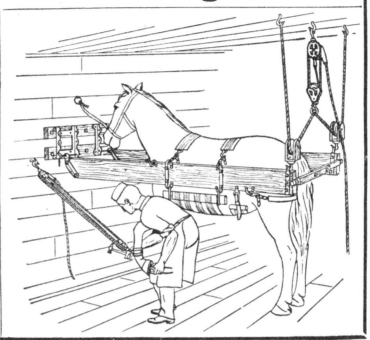

Strong and **quick** to operate.

Perfect **safety** both to horse and man.

Hundreds in use now.

Any foot, hind or fore, can be handled with equal **ease**, and by an automatic device.

Write us for circular which will tell **you all** about this Rack.

Manufactured by

HUNT, HELM, FERRIS & CO.
HARVARD, ILL.

BARCUS Horse Stocks

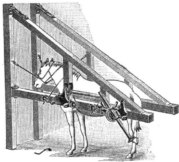

The most easily operated; the simplest and most perfect in every detail. Perfectly safe to both horse and

A perfect automatic device to hold any foot in any position perfectly solid. Not the cheapest but the best.

Write for particulars.

GEO. BARCUS & CO. Wabash, Ind., U.S.A.

CANADIAN HORSE STOCK CO.
Hamilton, Ont., Can.

Old Ads for Shoeing Stocks

Here's an old photo of a professional ferrier pressing a hot shoe to the hoof. Patience is the by-word, but certainty is important as well. As Lungwitz states on the previous page, sometimes the best way to approach a skittish horse is quickly and cleanly, as if you've been doing it every day for a month. That doesn't mean you should throw caution to the wind. It means, if you haven't got the skill and demeanor to do it, pay a highly regarded professional and get ready to watch and learn.

The legendary Morris Elverude of Oregon, a great teacher, teamster and human being.

Chapter Three

MENTORS

"Have something to say about the direction that horse is going."
- Lynn Miller

How We Best Learn

Working horses is not a science; it is a craft. It's a craft chock full of subtleties. But even so, it's a craft which depends on a lot of important mechanics. There's all the business about the harness; and how it fits, how it works when it's hooked up to the eveners, how all of that then hooks up to whatever it is you are pulling, and whatever work you're trying to do at the back end of the horses. That geometry is important, but if working horses is to become an art, that geometry has to be married to a natural understanding of all of the vagaries of the teamster's craft.

When you are first getting started working horses, for some people (people like me) the biggest single hazard comes when they/we just pretend to know what they/we don't know. It looks so easy, watching old guys do it, old guys who've done it for a lifetime. It doesn't look like there is that much to know. But when you pretend to know what you don't know, and you carry that arrogance into this craft that is so subtle and so rich with all the attendant hidden inter-relationships and important juxtapositions, you put yourself at risk and you put the animals at risk. My very first experience working horses, that runaway with Goldie and Queenie, taught me that; and the lesson has served me well my entire long life.

In my first two years with the team of Belgian mares, I spent a lot of time watching, listening to and absorbing from the teamsters I could find. I went as spectator to plowing matches, county fairs, the state fair, and living history recreations like the then new Dufur Threshing Bee. I tried to join a Draft Horse Club and found out that back in the early seventies you didn't just sign up, you had to be sponsored. I struck up a friendship with a cigar-chewing Dane named Ray Drongesen who lived and horse-farmed in Harrisburg, Oregon. He was easy to talk to and seemed happy to have me around. If someone new and young came along, thirsty to learn about work horses, he was drawn to them out of a natural need to be helpful. He was unique that way, it was more common to get the cold shoulder from older teamsters.

Ray and Don Degas
We might learn from mentors by listening to their instructions or observations, we also learn by their examples.

Back in the mid-seventies, while Ray and I were working together, he got a call from a local man who said, "I've got a big black stallion. You can have him if you can catch him." Ray had heard about the horse, a registered Percheron name of Don Degas. Seems the gentleman had bought the colt and put hm in a pasture and, one thing after another, it grew up and got mean. With the tattered remains of an old halter round his head, that stallion did whatever he liked in his small pasture, because every time the man went in to try to catch him he would race up striking and trying to bite. Here was a case of a horse no man owned. I went with Ray and watched. The man set in babbling to caution Ray. He politely interupted and said, "don't want to hear that right now, it's time for me to catch this horse." Ray, with halter and rope in left hand went through the gate as casual as you please. The stud snorted and came racing up. Ray looked the other way nonplussed. The horse hesitated, broke his run and walked up to Ray, ears back and snorting. With a smooth motion, Ray took hold of the remains of the old halter that hung round the stud's nose, and pulled his head down and towards Ray's chest. Ray stood still and looked at the horse's face without making eye contact, holding him secure. The stud did little sideways jerks that subsided, then he stopped resisting. It took a matter of seconds. Without moving his position, relative to the horse, Ray slipped the lead rope round the stud's neck and took the old halter off. He then put the new halter on, smooth as if it happened everyday. He stroked the horse's neck and proceeded to lead him forward. The horse baulked and made a motion to jerk away. Ray pulled his head sideways and put himself right back up against that horse's face, holding his head down. After a minute or two, he lead that horse out of the pasture. We put a long rope around his back like a brichen and together we gently coaxed him up the ramp into Ray's old 1950 Chevrolet stock truck. Later driving away, I asked him "how did you do that?"

"Lucky I guess", was his answer. "That Don's one smart horse, he can tell if you are afraid. And he took the time to measure me up. When you stand up against them and hold their head down so's they can move to get a focus on you, their eyes make it seem like you are twice as tall and that there are three of you, one at each eye and one at the middle. Trick is not to turn the tables on 'em. Don't let them get scared." Ray took Don home and within a week had him driving single, within a month he was working him in three abreast with King and Ruby on a plow. He bred mares with him and worked him right up to the end.

For me this story is one of both horsemanship and economics, it's your job to figure out how so. LRM

I met Jiggs Kinney, of Columbus Junction, Iowa, in Cedar Rapids in the late seventies. I bought a stud colt from him in the early eighties. There were stories of this man which may have felt like the yeast of legend, but none of those stories ever diminished the truth of his stature and influence in draft horse circles. He raised the best of Belgian horses, as many a showman and breeder did attest, and he farmed his fields vigorously with them. He was a true magician in the field; there was no way to "see" what tricks and devices he might have used. Those horses, they worked for him as he did for them.

When it comes to the inexhaustible subtleties of true craftsmanship, getting to some level of mastery draws upon and depends on the examples of so many who've gone before. Many were unsung heros. They did their work off to the side, out of the direct light and often without aknowledgement. They didn't expect or ask for thanks because they weren't there to be a mentor or teacher. They were there because they loved the teamster's craft, they loved the workings of draft animals. But it must be said, and often, that they are the ones who kept this way of working alive and inheritable.

Putting the secrets into a book that will rest on a shelf, that exercise has merit. But this information is seed-like and needs to be planted by people who will actually use it.

Toughest customer of working horses was quiet, steadfast Dan Kintz of Mount Angel, Oregon. By example he taught countless newcomers to the craft he so obviously loved. He taught me the effectiveness of the perfect whisper.

Late in his life, Jess Ross became a dear friend and partner in the fields. We first met in the Big Hole country of Montana where he was working buckrakes in the hayfields. He was the quintessential cowboy/teamster who quite literally laughed in the face of any danger. We did not agree on this, but there was no denying his ability to make it work; he loved to be on the lines when a horse or team "frightened themselves" (his words) to run away. He would make a certain effort to stop them; if that didn't work he'd smile big as all get out, chuckle, and urge them on to their own earned exhaustion, then whisper to them til they joined him completely and for all time.

Long ago, on one of our trips to the big annual draft horse auction event in Waverly, Iowa, we met a tall, laconic gentleman named Everett Hildebrandt. He impressed us greatly with his close-in, passionate skills training horses. On this occasion he had consigned a young, magnificent Belgian mare to the sale, as he had often done before, and folks were talking as how they didn't want to miss when that animal came in the ring. Arnold Hexom, the auctioneer, would make sure there was enough time for Everett to demonstrate what that mare could do, and what he had done with her. There was a foredrawn conclusion that whatever consignment Everett brought would be a high selling entry. In the crowded sale ring, noisy PA system blaring, Everett would drive in his filly and proceed to remove the lines and quietly talk her through turns, back-ups, and side steps. She calm at every step, he beaming with justified pride, prospective buyers salivating. The eventual price would be thousands of dollars, well above most every other consignment.

As Hildebrant explained to me, "the secret is that you carefully purchase an intelligent weanling filly of the best bloodlines and perfect conformation. Then you take her home and let her grow up while you quietly and patiently train her to do whatever you want. That simple. After a couple of years of selling that sort you get a reputation and people will pay whatever price for her."

"But why," I asked, "after all that work and success, would you then turn right around and sell that superb youngster to the highest bidder? I'd want to keep her on the ranch, raise her foals, work her."

"Because", he answered, "this is what I do. I did my part and I'm always anxious to do it again, to find that next one. Time to move on."

I think it was the same trip when I went to visit Herschel Griffith's horse barn. He had an eight horse hitch, I had purchased his lead team, Bob and Bud, which are pictured on the front cover of The **Work Horse Handbook**. Herschel's barn was a tad bit unusual; he had ropes of colored pennants (like those you see on a used car lot) tied around the ceiling and blowing in the breeze of the fan. He also used a tape-player on which he had the soundtrack of a horse show complete with organ music, announcer banter and crowd clapping. None of that bothered them, through all of that his horses chomped quietly on their hay.

"Giving them all of this as background just means when I go in the ring there is that much less for them to concern themselves with," he offered.

(Horses are amazing and will condition themselves to incredible environments. Think about those mounts the police use in Manhattan, standing quietly below skyscrpers with flashing lights, and thousands upon thousands of anxious people.)

You might have to look out back, or follow a lead down a long dusty road, but the hills are full of teamsters quietly living their lives working in the manner they choose. Many of them have helped hundreds of people get started, and without much herald. Chuck Baley, above, and the late Faye Pishon, are two such examples. I had the distinct pleasure of teaching a workshop with Chuck down at Fort Lewis in the four corners region. He's the real deal and a secret resource for lots of folks who get more publicity but have less skill.

George Speisshart in the early 70's.

At the State Fair, I watched the eight horse hitches compete and found myself fascinated by George Speisshart and his Percheron hitch. I tried to make friends with him but he was highly suspicious of anyone, especially young people, who were out to "steal" his secrets. Every question I asked was a "dumb" question and he let me know it. He seemed to say that if you weren't born to the teamstering business, you had no way in. He finally got rid of my pestering. He waited until there was a crowd of draft horse folks around his stall display at the Oregon State Fair. Then he pointed to me and told those present one of my lame questions and laughed, inviting people to join in the ridicule. I walked off, never to return to his camp, though for all the time I knew of him I couldn't help but admire his exceptional horsemanship.

That was back in 1971, things were very different then compared to now. It took me awhile to realize that what I was trying to get in on was a circumstantial secret society. At that point in North American history, this was a culture of hold-over teamsters. Our culture, in rural communities, had changed dramatically during and immediately following World War Two and working horses in harness was considered one of those things of the past. People were regularly ridiculed for hanging onto this lost craft. Yet, at the same time, those teamsters were fiercely proud of what they knew how to do, and they were holding on to this in the only ways they understood. They became secretive and protective. It was difficult for them when young people came in assuming that this here craft was no big deal, people like me who came in pretending to know what they didn't know, and allowing as how they could just take a pair of horses and harness them up and go do something like it was easy. Because when I acted that way, when I put myself forward that way, I was actually demeaning them. I was diminishing them and their value as teamsters, their value as guardians of the art.

Dan Kintz 1975

There was a tall, sinewy fellow of ruddy complexion whose smile seemed to be turned inwards; his name was Dan Kintz and he was a horselogger. He had middleweight Belgians and besides the forest work, he did his farming, plowing forty acres or more each year with a team and walking plow. In those early years he also participated at the state fair showing in the farm team class and the pulling match. Dan was the silent type.

I attended a state plowing match, back before I ever plowed with my own horses. Dan Kintz, a Champion plowman, had two Belgians on an Oliver 40 walking plow. The land-side horse was a long two-year old stud named Prince. Dan was resting at the headland, his team standing at the ready when Prince side-stepped over the tug, putting the chain between his legs. Dan, cigarette hanging from his lips, lines in his right hand stepped up and placed his left hand on Prince's lower hip and asked him, quiet-like, to step back over. Prince crow-hopped in place and cow-kicked Dan knocking him down. He hit Dan in the upper leg and it must have hurt. Dan got up, no change of expression, and went right back to the same spot, hand on the stud's hip, leaning into him while saying "step over son". The colt tensed up, then relaxed and stepped over. Dan stroked him and returned to behind the plow handles.

I learned from Dan just watching him. I learned about steady persistent expectation. I learned about where to stand, how to stand, and that the safety and hazard zones have to be felt.

Ferd Mantei in 1974.

Ferd Mantei had a big, black team of draft horses. To the casual observer, it was hard to tell the three of them apart, not because they looked alike but because they stood, moved, gazed about, worked, ate, absorbed, and participated in perfect synchronicity. I remember at his farm, where a club plowing match was being held, Ferd allowed that all contestants could tie up their horses and mules all along the inside of the barnlot fence. Mantei had a big, old, classic, gambrell-roof, split-level dairy barn which formed, on its long side, at the lower-level loafing shed, one perimeter of that barn lot. It was a bank barn; the lower level on one side, the second level on the uphill side. On the lower side, center of the barn was a man-size door from which a heavily reinforced 3 foot wide ramp dropped 10 feet down to the barnlot level. In the morning, as teamsters arrived, unloaded animals and proceeded to tie up along the lower side fence, I saw Ferd amble up that ramp and disappear into the bowels of the barn.

A little time passed and I spent it watching the activities of contestants and animals preparing to go to the field. They all stopped at one point and turned to look at the barn. I did as well. Here came Ferd, walking down the steep, cleated ramp. Right behind him was a big black gelding in harness, bridled, no lead rope. The man and horse were matched for step. Soon I saw that a second black gelding was following the first, same pace. Ferd, without looking back, without a word walked to the middle of that large barn lot, both horses in a line behind him. Ferd stopped and turned. The first gelding stopped and very casually looked around at all the new animals in his pen. Ferd took two three steps to the right, offset now from the head of the first gelding. When the man stopped the second horse came straight around the first and stood in perfect position side by side with his team mate. Ferd nodded his head and walked away from the quiet standing pair, going to each contestant and welcoming him or her to his farm and the plowing match. The black team remained in the center of the lot.

Finished socializing, Ferd went back to the team and setup the driving lines and proceeded to drive them out to his waiting plow. I didn't know what to think of what I had just witnessed, partly because I was young. I didn't know the depth of the relationship possible with good draft animals.

At noon that day the contestants unhitched from the plows and returned to that barn lot for a break. Ferd was the last one into the lot, ground driving his black pair to that exact same spot. Once again they stood absolutely still while he did up the lines, separating the horses. Then stepping to the side he nodded once while facing the near-side gelding, the horse stepped ahead and proceeded to that steep ramp. When the big black was halfway to the ramp, Ferd nodded to the teammate and the horse proceeded, walking quietly, to follow his work mate up the ramp.

I couldn't hold on any longer. I had to know, so I walked up alongside Ferd as he followed his team, "Mr. Mantei, how did you get them to do that?"

"Do what?"

"Get them to stand like that and then walk without a lead rope and go when you wanted them to?" I asked

"If you're lucky enough, I expect you'll learn that some day."

It made me mad, his answer. So I went over to another horsemen I knew, Monte Rumgay, and I asked him the same questions. Monte said, "He answered your question you know. He answered when he said 'I expect…' Once you learn about horses then it becomes about your expectations. If you think it can't be done then it probably can't. But if you expect it and are ready to insist, well… There's another part too, though. You've got to have it under your skin, Ferd's got it under his skin; he's one of the best."

Here are three dear friends and great teamsters, gone now but never forgotten. Bud Dimick (top left) Charlie Jensen (top right) and Bud Evers (below).

SUGGESTIONS: When you ask questions make them very specific and answerable. Example: how long should the doubletree be for a horsedrawn mower? or What is the first thing I hook up when hitching to a tongue? Stay away from general questions. FAR AWAY. Never ask questions like: Can a woman work horses? or Do you think that I could maybe farm, like, say 120 acres in the mountains with work horses after work and on the weekends, and should I keep my small children away from them, and what if my wife is afraid of them?

With working horses and mules, I've seen so many styles, but what has made the strongest impression on me over the years has been the *manner* each individual brings to the work.

Mike Adkins (just below) gets in real close with his animals, and lays down a constant manner and banter of serious encouragement. "Now boys, I know you can do this. And you know you can do this, so lets just stay focused and git her done. No foolishness. Right there, that's what I mean. Come on back. I'm not going anywhere, we're doing this together." All of it with an easy southern drawl that makes puppies offer him their empty kibble dishes to lick. Mike and I have a running competition: he likes to say that he knows he's the best looking teamster in the world. I have a hard time arguing with him as to looks, but I do make the case that he can't come close to me when it comes to bragging on who is the best 'equininian' raconteur. Mike's got better animals, a better way with women, and a voice like warm sour milk laced with cayenne pepper and chocolate, but he's scared of a well-placed exaggeration or two. So at least I've got him there.

Two PNW teachers, John Erskine and Walt Bernard. John has been helping folks for decades. Walt is one of the new breed of passionate instructors.

Lise Hubbe (left). Though she has an impressive list of mentors, everyone a giant in the business, (people like Dick Brown, Harry Lehman and Dan Kintz) to name but a few, to my way of seeing it she taught herself to work horses. Her determination and calm, steadfast work ethic kept her in the game til she mastered it. She gets ornery with me when I say that, quick to shift the focus away from herself and to acknowledge these men who helped her along the way.

Marvin Brisk disappears into each animal. Here he is using a swingle-tree to pull the barn's haul-back rope for hay. Photo by Joel Sokoloff

Joyce Sharp is a sterling example among of Pacific Northwest teamsters, a capable and wonderful human being. Photo by Kristi Gilman-Miller

Herman Daniel was so happy to be alive and working horses, that his teammates figured they had arrived in a pretty darn good setup. Portrait drawings by the author.

Forrest Davis was tough, old-school, and 'by golly knew the way to get it done'.

Joe Van Dyke selected horses ahead of the game. He wanted those which he could make look better and which made him look smooth.

If mentor by definition is someone who imparts wisdom than certainly a teamster *as example* fits the definition. The list goes on long as all of our arms connected; from Paul Birdsall to Jack Carver, from Donna Anderson to Leo Silva, from Gene Hilty to Jimmy Klein, from Luke Vastine to Mike McCormick, from Abe Yoder to Gary Eagle, from Scott Link to Judson Schrick, from Doc Mustard to John Erskine, from Wilbur Hilgedick to Vernon Madsen, from Mel Anderson to Willard Wilder, from Fred Baker to Harold McMain, from Aden Freeman to Jim Cornish, from Donn Hewes to Cornelius Verduin, from Jacob Yoder

Robert Clark of Montana was a man who influenced people with his daring enthusiasm.

to Robert Yoder, from Don McInnis to Andy Meyerhofer, from Jay Bailey to Charlie Orme and far, far beyond.

That's a very small piece of a list of folks; you can add a whole bunch more from the info and captions in this book. And when you're done with that you can start building a list of your own.

So you think you want to learn this craft?. First off; I'm telling you that you can do it. But second; I am cautioning you that this is a way of working that can get you, and the animals, in a whole lot of trouble, fast. You don't know what you are doing. You need mentors. Notice I put that in the plural. A good mentor can be most excellent but two or more mentors give you a far better sense of how various this craft is.

Montana's Doc Hammill, (left) veterinarian, western historian and draft horse instructor par excellence. Photo by Kristi Gilman-Miller

(Left to right) Jim Butcher, Kenny Russell, Lynn Miller and Mike Atkins.

"Regard" is earned and wants to be more generally valued.

These teamsters are well regarded for good reason, all tests passed.

Bob, "Bulldog", Frasier was a horselogger and farmer from the Noxon, Montana area. Nothing frightened him, especially not pretty ladies who wanted to learn about working horses.

Ken Geiss mowing with his miniature horses. Ken teaches by example and by sharing his myriad inventions through the pages of Small Farmer's Journal.

Oregon's Roger Dougherty and his superb Percherons on the plow. Some people mentor by example. Photo by Dick Brown.

Ben is giving Dad, Ike Bay, needed supervision some years back.

Montana's Charlie Yearian.

Don Lee plowing with his Percheron pulling and logging team at Yamhill competition. Photo by Kristi Gilman-Miller.

Above: Mac McIntosh, of Terrebonne, Oregon, patriarch of the McIntosh clan, long-time devotees of good Belgian horses. Below: Mike McIntosh (Mac's son) is superintendant of the local school district and a farmer, teacher, family man and friend. Photos by Kristi Gilman-Miller.

A scene at the Dufur Threshing Bee, back when horses were central, with Mt Hood in the background. Mike McIntosh drives his four abreast on the header while Jeb Michaelson drives his greys on the header-box wagon, receiving the harvested grain. Photo by Kristi Gilman-Miller.

The McIntosh family threads their working horse pursuits within and throughout their entire lives. Each late summer they do their own threshing bee. Joanna McIntosh took this picture, a couple years back, of the volunteer crew.

When the Dufur Threshing Bee was going full force, the McIntosh clan provided a big share of the demonstrating horsepower. Mike drove the header, kind of a grounded version of a cross-between helicopter and portable escalator. To turn the Rube-Goldbergian grain header, Mike, driving two separate teams, must back the right team while fanning the left horses and asking them to walk around. Photos on these pages by Kristi Gilman-Miller.

It's possible for Mike to turn this giant machine right in place. When the corner is made, he swings that left team back around to form the four abreast and proceeds to cut and elevate more grain into the waiting header-box wagon. It goes against the horse's nature, but these beauties walk forward with the mowing, elevating and paddle-wheeling all going on right in front of their noses.

Janelle, Mike's daughter, on grain binder.

James, Janelle and Jacob (all siblings) are incredibly fortunate to have been brought up in the very midst of everything horsedrawn. All of us who are permitted to feel like members of that extended family are blessed. Photos by Kristi Gilman-Miller

Jacob drives his Percheon team on the stage coach and James, below, drives the carriage, both for sister Janelle's recent marriage. All photos these pages by Kristi Gilman-Miller.

"Marry me and I'll never look at another horse!" - Groucho Marx

*Above; Mike McIntosh by Kristi Gilman-Miller.
Below Mike and Lynn Miller plowing together. Photo by Bing Bingham.*

"Ride the horse in the direction that it's going" - Werner Erhard

Oregon's Schaun McInnis, photo by Kristi Gilman-Miller.

Chapter Four

repetition

'You can lead a horse to water, but a pencil must be led.' - Stan Laurel

When I try to stay on a beneficial routine, fold it into my daily life - make of it a habit, I have never been able to get past whatever new urgency presents itself, urgencies that strip away the validation for beneficial routine. My brain knows that I need to do this thing once a day for fifteen minutes, or whatever the required framework, and that the repetition is as critically important as the months I will be engaged in the routine. But then along comes some unforeseen demand, for example an injured horse that requires of me that I find half an hour every day to change its bandage and doctor the wound. Now that original 'beneficial' routine must be set aside, and the primacy is lost. If I didn't change the bandage there were consequences, the urgency was genuine. If I stopped that original benefical routine of bringing in the young team to the tie stall for grooming and

Dick Brown, photo by Robert Mischka. Getting big hitches to the field for a day's work and then back to the barn again requires efficient routines, most especially if it is your need to do this day after day. Every minute saved is almost as important as making sure that the working horses have the best feed in, and with, the most comfortable of stabling. Working horses like Dick Browns are made to look this good, they don't just happen. So what you see here should have a far more important impact on you than a simple diagram of a suggested stabling or a recipe for feeding. Because one thing I have learned over a lifetime of working horses is that clear goals and best intentions manifested your own way, that's the ticket to success, not a recipe. The picture you see above is the culmination, a masters true communion.

feed, the consequences were soft and invisible. So, for me, a slave to the moment, the primacy is easily lost, I guess that might make me weaker than most, but there it is. Thankfully circumstance fixed that.

The very best results with working horses wants routine, wants repetition, wants the long and tedious. My successes came fast and true when I found myself locked into a routine I could not vary because if I did I would lose the farm.

When I finally took the plunge and the gamble and bought my own first farm I had very little money to get me started and on through that first growing season. I had some farming experience, I had a team of Belgian mares that I drove for fun, and I had the cloud of a challenging commitment. I had contracted my down payment on that farm. I needed that first growing season to succeed well or my adventure was over. I took a chance and used a small amount of the money I had to purchase a seasoned (say 'getting old') team of geldings to add to my work string, Bud and Dick. They were bay Shire Percheron crosses. That meant I had four head of work horses and no tractor. And I had fourteen acres of vegetables to plant and twenty acres of grain to plant and twenty acres of hay to put up. That meant I had to design a routine to get it all going and maintained. If I did not stay on target and get the work done, day in and day out, all would be lost.

Quickly I found that time, valuable time, could be saved by designing the day's working armature to streamline the repetitions. My farm had been a dairy and it had a loafing shed which adjoined the barn which in turn matched up with the milking parlour. I built a row of tie stalls sufficient at first for four head, then growing to eight. That was because Ray Drongesen and Charlie Jensen started to bring their teams over, on occasion, to help me out. The first stalls I built were 5 feet wide by eight foot deep, sufficient for one horse. At the head was a hay manger and grain box. Within a short period of time I learned, from Ray, that having a double stall, 10 feet wide, would save me a lot of time as I could harness a team, rig the lines and back out to ground drive to the implements. Before that discovery and its time saving comfort, I had led the horses outside and tied them to rings I had mounted on the barn wall,

Dick Brown, photo by Robert Mischka. (Note horse with the mane is stallion). Hitching or unhitching: to most it is unfathomable that one person could walk within six, eight or ten horses and arrange them with the air full of nothing but willing cooperation. This is impossible for one who has no experience, This is incredibly dangerous for one who is separate from the animals. This is not a carnival trick. This view is from within the less than attractive housekeeping portions of efficient horsepower but it is OF the magic that is possible, the usable, soul-defining magic.

My good friend and exceptional lifetime teamster Kenny Russell of Poplarville, Mississippi. Kenny and his wife Renee have given many people excellent beginnings through their many Work Horse Workshops. Photo by Kristi Gilman-Miller

this while I finished harnessing and rigging the lines.

The added convenience of the double stall meant that I could, when breaking from the field work, ground-drive the team into the barn and right into their stall where they would welcome the spot for a rest. Once there I could take my time pulling bridles, lines and unharnessing while the horses chomped on their hay. I once figured that the double stall saved each of my field work days at least 15 minutes. Adding that up, it meant that the 45 days I spent plowing, harrowing, planting, and cultivating in the spring resulted in almost 11 and a half hours saved. To my work fevered brain that meant I could, that season, plow up another acre and a half with the saved time!

All of that, of course, is but one way any individual teamster might customise the mechanics of working horses for the economic benefits, but it doesn't speak to the magic. That came from the routine. In those early years, every day, six days a week, I took my four horses into their stalls, fed them - groomed them - harnessed them and then took them out to the field for long hard days of work ending with a return to those stalls. Setting that actual working part aside for a moment, the stabling came, in my mind, to be the most important part of the whole routine for it taught my horses and I so many things about what is possible in the relationship.

In the first days, I would go out to the pasture and halter the horses, two at a time, and lead each team all the way to the barn and their stalls. Soon I figured out that having them stay in the loafing shed at night meant they would be at the gate waiting to come into the stalls. Goldie, Queenie, Bud and Dick each had their own positions in the two double stalls. I would go to the loafing shed gate and halter the four head, then carefully lead one horse at a time through the gate, holding back the anxious remaining animals. One day, after a couple of weeks of doing this, I went to the gate and unlocked it and noticed across the barn that my pigs had gotten out of their pen. I trotted over to put them back. When I returned to the stable area all the horses were in their 'proper' stall positions, without halters, standing quietly eating hay. They had pushed through the gate. I was angry but only for a short instant. I went into the stalls and haltered each of them easily. From that day on, making sure each manger had hay in it (no grain), I would open the gate and stand back to watch 'my' horses casually walk to their stall positions. It was magic. Simple magic. And it was born of routine.

After this was 'learned', I found that I could turn the horses out on pasture at night, then in the early morning I could go and open the lane gate, the one jersey cattle had used for years to enter the loafing shed, and the horses would come of their own free will to the loafing shed to wait for the barn gate to be opened. At that point they would go to their stall positions to be haltered, fed, groomed and harnessed - all before breakfast. One of my most satisfying repeated experiences was returning to the barn after a bowl of cereal, or of an evening after supper, and soaking in the view of 'my' horses all standing in their stalls harnessed and quietly chomping on hay. That line-up of brichen-encircled hind ends still fills me with an "all's right in the world" feeling.

Kenny Russell, of Poplarville, Mississippi, tells the tale of his father, who insisted on harnessing a team every single day, whether there was work for them or not. Kenny spoke of going out to the barn as a young boy to see his father with his arms around the backs of his team. Powerful magic in that.

Asking of work animals that they stand comfortably, tied to a certain position, quiet for long periods of time - from half an hour to an hour and a half (or even sometimes longer) - is a large request of these grand beasts. Its not a natural thing for them. They want to be heads free and able to take off, for self presevation, at the moments notice. We can make it easy for ourselves and pleasureable for them by offering that this restrained and stabled time hold the

LaDene Scott of Dorena, Oregon taking a break during a plowing competition. Photo by Kristi Gilman-Miller

Ryan Foxley spreading manure on Littlefield Farm in Washington. Photo by Joe D. Finnerty. Ryan began his pursuit of the teamster's craft as a young adult, and over many years has become a true master now sharing his experiences through his music and his writings.

moments that they eat hay and grain. Doesn't take but a day or two for them to actually want to be in that stall.

And while they are there, we brush and comb them and untangle their tails and manes. We fuss with whatever needs fussing with. We mumble and talk, and whistle and breath and they learn to be close to us, and that we are to be accepted inside their 'circle of being'. Add to this the harnessing, and adjusting, and later the unharnessing and wiping down and you have a daily 'beneficial routine' that can make of these animals willing partners.

I recall watching a juggler on a TV variety show, spinning plates on the ends of sticks, many of them - even one on a short stick balanced on his nose. It was magical. Pure magic. You could see what he was doing, no strings no gadgets no tricks, all balance. I realized that magic requires balance and balancing.

The stabling and harnessing routine is similar, no strings, no gadgets, no tricks, just a person going about a set of preparations. The results we see as magical but truly we only see them as magical if we allow that vision or view. It is possible to have these things happen daily and for certain people not to see the results as anything special or to see a very different picture, perhaps fortune ahead of magic.

I once visited a Minnesota farmer who had a big hitch of Belgians. It was not a fun trip. A year before I had bought a pair of registered mares (in the late 70's) from the Cedar Rapids, Iowa sale. Took them, and others home, and that pair of open mares, Bobbie and Carol, quickly became a favorite. My friend and helper, Herman Daniel, and I used those girls right away for field work. A few months later while mowing we figured out that Bobbie was pregnant. I called the previous owner on the phone and asked him if he had exposed her to a purebred Belgian stallion and if so could I get a breeding certificate from him. He was angry and indignant, he said that was impossible as she was a barren mare. He then asked how I was getting along with them and I said "wonderful, Herman's out mowing with them right now." He called me a liar. Said the team was nothing but a pair of barren cranky old mares that ran away every time he hitched them. After that he refused to talk to me.

After months of negotiations I got some blank breeding forms from the association and, on another draft horse buying trip east, surprised this man on his farm. He acted as though I was some long lost friend. Seems that he had seen pictures, in the interim, of me working those mares and he wanted to know what secrets I, young as I was, had up my sleeve. It was easy to get him to provide the info and signature I needed. Then he insisted I go out to the barn with him to see his horses. As we approached the big old dairy barn, with a gradual ramp up to the open double doors, I could see the beginning of a long line of tie stalls and the first animal - a stallion, one back leg cocked at ease, was standing well back in the stall.

Bob Chichton of Massachusetts with his three mule team competing at the United States Draft Horse and Mule National Plowing Championship at Carriage Hill Farm in Dayton, Ohio. Photo by the author.

My host muttered something under his breath and told me to stay back. I saw that adjoining the same wall as the stallion, on the outside, was a empty oil drum, a long handled broom and a length of pipe.

The man went wide to that side so that no horses could see him and he snuck up on that oil drum. Pipe in hand he swung hard and hit that drum. The stallion, I could see, jumped up into his stall and I could hear what sounded like several other horses doing likewize. Then the man traded the pipe for the broom and, still out of sight, passed it by the edge of the door. Immediately the stud's back legs shot out in a killing kick. "Okay" he said, we can go in now. Inside the barn were twenty head of beautiful Belgian horses, everyone shaking with nerves. Here I was experiencing a case where the stabling was likely part of a nasty nightmare for those good horses. Didn't surpise me that my mares ran away from him, nor that Bobbie never settled from breeding. This man had no balance and no vision.

If you can find how to 'balance' a kind routine and then discover where to look for it, the magic is yours for the amplification.

Adjusting the plow at the U.S. Plowing Championship in Dayton, Ohio. Photo by the author.

Chapter Five

"Those horses must have been Spanish jennets, born of mares mated with a zephyr; for they went as swiftly as the wind, and the moon, which had risen at our departure to give us light, rolled through the sky like a wheel detached from its carriage..."
   - Théophile Gautier, Clarimonde

make 'em walk you damned fool!

Charlie Jensen, back when he was a fixture in my early work horse years, kept a four-up of old-style "chunk" grade Percherons. We called them black, but technically they were the darkest of bays in color. Charlie loved to plow with three head and a riding unit (sulky plow). And he enjoyed driving a four-up in parades and shows. He was so good at it that you might not see the skill, he was invisibly good. After I got going with my own farm and had Ray Drongesen helping me, it was Ray who talked Charlie into joining us. I had twenty acres to plow and didn't know what I was doing. Ray and Charlie were anxious to tune up for the two spring plowing matches. While Ray and I got along well, Charlie made no bones about the fact that he thought I was a dumb kid.

Only the second time I had ever plowed and Charlie was there by my side with his three abreast. I had Bud and Dick hitched to Ray's riding plow. The day before Charlie and Ray had opened the field with two "lands" running parallel. Ray was off that day, so Charlie went to one land and I to the other, we were plowing maybe fifty feet apart. My team had done some time in Nevada as pulling horses, so when the plow was in the ground they wanted to really lean into it. They made the difficult job seem easy, I was having a ball with that wet spring soil rolling over so sweetly. The horses were moving along at a good clip when I heard Charlie yelling at me "Make 'em walk you damned fool, make them walk!" I pulled the team back and now I had to hold them pretty hard because they wanted to go faster. But Charlie made his point. He glared at me the entire rest of the day. His own horses took each step deliberately, leaning into the collars, and moving as though at some choreographed slow motion walk. Later I would learn all of the hazards of plowing at a "clip" none the least of which was that the horses would play out long before the second hour.

Tough, tough Charlie Jensen, late of Harrisburg, Oregon. At one time in his life he was repairing a dump truck, the hoist was up and he was wrenching and didn't notice the very slow descent of the box, hydraulics leaking. Before he could get his head out it was pinned between the frame and the box. Took a few minutes to get him out of there. He claimed to me that before that accident he couldn't think as clearly as after. Fine, tough man, great horseman. He's gone now but I can still hear him yelling at me from across the land, "make 'em walk you damn fool!"

1975, Ray Drongesen plowing with my horse, Dick, and two of his Percheron colts on my farm in Junction City, Oregon.

I had been plowing just four days when the gentlemen left me all alone to my work. And I was feeling pretty cocky when a curious thing happened...

But first a note to set the scene: The sight of a big team of horses or mules in harness, out working in the field or hooked to a wagon going down the road, is a powerful attraction for folks. A lot of farmers and ranchers have used it to good commercial advantage, powerful commercial advantage. I know of quite a few good examples spread all over North America where folks have added tens of thousands of dollars to their bottom line by incorporating working teams of horses and mules. There is something interesting about this draw that folks feel towards the working team. It's like all folks seem to have taken ownership of the idea of working horses. I remember a particular experience I had early on when I was just getting started.

Once again I was working that team of geldings, Bud and Dick and I was plowing at my farm in Junction City which was right on the city limits. The east side of the farm was bordered by a county road, a two-lane paved road, no shoulder to that road. There was a ditch there on the edge of the pavement and then it came up to a berm and there was a woven and barbed wire fence. Just over that fence I had a little piece of hay ground and then two hundred feet across to another fence and over to that ten acres that I was plowing to plant into vegetable crops. I was using Bud and Dick and we were plowing east to west and that county road was running north and south. As I was plowing, coming back east with a view of that road, I noticed a sedan pull up slow and then stop right there on the fence line, right in that traffic lane. And a woman jumped out of that car and ran around the other side of her car and snatched a little boy off the passenger seat and walked over to the fence and she was pointing at me and talking to the boy.

I was used to people gawking; paying a lot of attention to the horses working. And cars would come up behind her and honk and go on around and she just left her car there. Soon she set that boy over my fence and then she herself proceeded to climb over the fence. About the time she made it over the fence I was finishing that west to east path and I stopped my horses and let them blow and here she came trotting, carrying that little boy, at a slow trot across that grass towards me. She had yet another fence to cross.

I thought, "Well she's probably coming to get directions or something, there's got to be something wrong with her car. Not much I can do, I'm going to stay out here and plow." And before I even got the chance to say "hello" she jammed that little two or three year old boy right up into my lap with his face right up against my chest and she pointed at the horses and she said, "take him for a ride."

Well when she did that it got Bud and Dick a little antsy so I just instinctively spoke to them and turned them around set the plow back into the ground, half thinking I was going to return that boy to the woman and tell her, "no thanks" but as I looked down this little boy was gripping my chest like a monkey, he was so scared and he was so excited. He would look over his shoulder at the horses as they were moving.

I don't know what possessed me, it was not a real good idea but I looked down at the little boy and said, "Would you like to go for a ride?" And I got a hesitant, slow nod. I could feel it on my chest more than I could see it. So I figured what the heck. I spoke to the horses and we headed on down the path, moving down that furrow, heading west, away from the anxious mom. That little boy would turn and look over his shoulder every so often at those horses walking along and then he would just jerk that face back around and bury it in my chest. I can only imagine what kind of an experience that was for him and now as an adult what he must think of that one little ride he took on that plow.

Well, we made it all the way to the other end of the field, pulled the plow up out of the ground and came around and set the plow in again and I thought I would let the horses rest and breathe for awhile, but I could see down at the other end that mom was waiting there with her arms crossed and she was tapping her fingers on her arm and she was tapping her foot. She was anxious and the car was still parked in the traffic lane. People were still honking and whipping around her car out there. She would look over her shoulder every once in awhile at the car and than back to look at us. So I went ahead and spoke to Bud and Dick and we returned down that furrow.

The little boy was pretty much in the same posture, gripping me like a monkey, face buried in my chest, every once in a while looking over his shoulder. I want to believe that he was excited but he was still genuinely scared. Before I'd even got the plow out of the ground, while we are still moving forward at the other end, that woman came and yanked that boy out of my chest and without a word she turned and stomped off to her car. Yep, she had taken ownership.

Ryan Foxley of Littlefield Farm, Washington, plowing with his Fjords. Photo by Joe D. Finnerty.

There it is, that moment when you stop to take a breath and they do too. Oregon's Wayne Beckwith and his mules, Fahta and Kerala. Photo by Kristi Gilman-Miller.

Canadian championship team plowing in International meet at Carriage Hill in Dayton, OH

Mike McIntosh uses an Oliver walking plow with his superb team of Belgians. Photo by Kristi Gilman-Miller. There are two prevalent ways that plowmen handle their driving handles. One way is to have the lines pass around the waist. Mike and I use the lines over the furrow-side shoulder and under the land-side armpit. There are two reasons for this. One is that if there is a problem, ducking the head and rolling the shoulders has the teamster free of being drug. The other reason is that with this system, turning the shoulders actually turns the horses ever so slightly. It does require that the lines are tied at precisely the right length so that the teamster stands straight up plowing.

Setting up six mules with rope and pulley hitch on Pioneer gang plow at Pennsylvania Horse Progress Days, photo by author.

Delani Hedley, Bart, 26 and Bonnie, 15, on 10" JD plow, Delani is 15 year old granddaughter of mule owner Neal McCool. Photo by Kristi Gilman-Miller

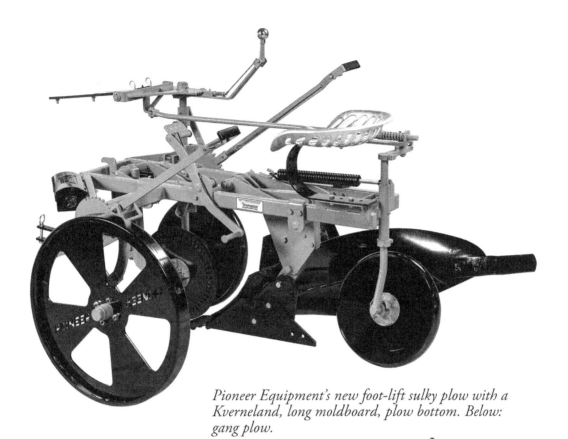

Pioneer Equipment's new foot-lift sulky plow with a Kverneland, long moldboard, plow bottom. Below: gang plow.

The subject of horsedrawn plows and their workings is vast and various. I authored a text titled ***Horsedrawn Plows and Plowing*** some years ago. It contains over a thousand illustrations and lots of information but in no way does it cover every make, model or variant. With this text I want to offer just a taste of the subject.

At the risk of over-simplifying: there are two classes of plows - the walking and the riding. You can guess their distinction in the names. The walking plow is a tool which shows great finesse in its precise design and equal finesse is required in its efficient use. The share, that portion of the plow which slices into the soil, features a point and a landside (up against the furrow wall), these employ a cupped design which creates suction to hold the plowing depth and the plow against the unplowed ground. Back behind, and attached to the share is the moldboard which curls the soil, crumbling its structure, while flipping it over and/or up on edge. The moldboard is affixed to the plow

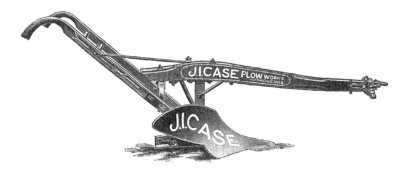

Walking plow above, beam-style riding plow below. Many different manufacturers advanced the engineering of plows rapidly from the 1880's on through to the 1920's. Millions of these implements were made and their solid construction gave them a very good chance of surviving to this date and in large numbers.

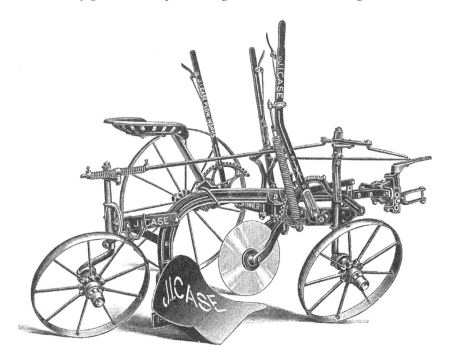

beam which sets angle (as per adjustable "angle of draft") and depth. Plow designs are a correct marriage of beam to moldboard. How well they work begins with the farmer/teamster correctly setting the hitch to the eveners and animals. Hitched too low for team and the plow wont stay in the ground. Hitched too high and the plow will run too deep.

The disc plow design allowed that stoney ground might be turned with a minimum of plow damage. This design with its high arching beam is quite unusual.

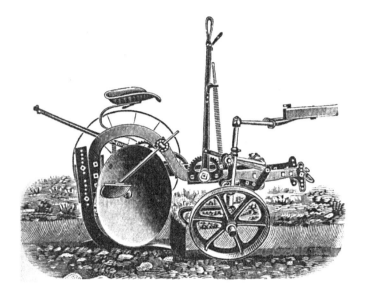

This is a bottom and backside view of a slat moldboard "gumbo" walking plow showing the bracing and landside construction. This style of moldboard allowed for more lateral disturbance as the soil passed, reducing sticking.

Angle of the Draft / Line of Draft

The correct hitch for a plow. The tugs are parallel to the line of draft from the center of resistance of the plow through the clevis pin to the point on the hames where the tugs fasten the hitch. (Courtesy Oliver Chilled Plow Works.)

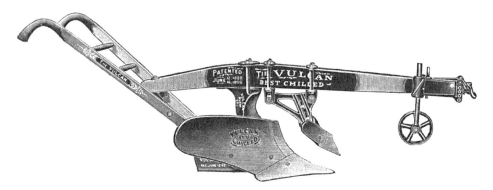

Above; Vulcan walking plow

Right; Slat bottom 'gumbo' plow

Below; Reversible hillside walking plow

Bottom; Three bottom walking plow

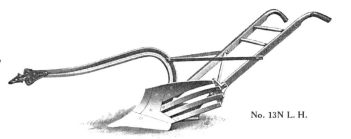

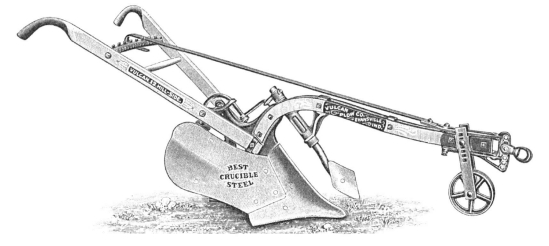

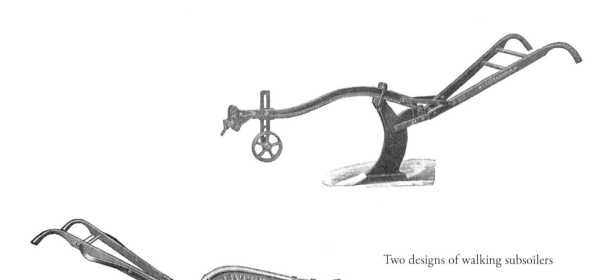

Two designs of walking subsoilers

Right and bottom right; Two views of a grape hoe or vineyard plow

Below; another backside closeup of a slat-bottom 'gumbo' plow

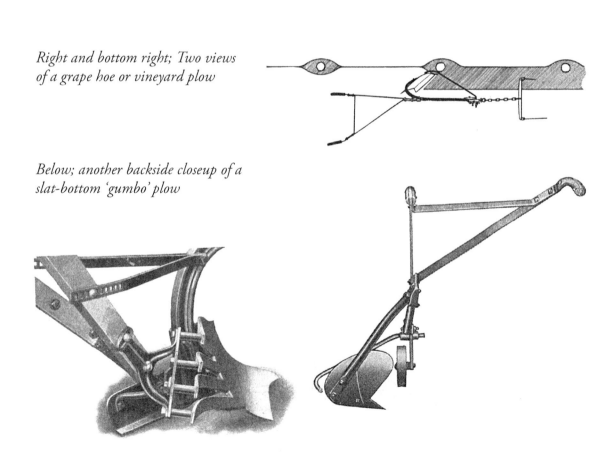

145

Oliver beam-style riding plow, which refers to the fact that everything hangs off the beam. Some call this style a glorified walking plow.

Oliver made some serious advances in its engineering of the beam-style riding plow including improved lifting quandrants for the handle mechanisms, higher trash clearance with arched beam, and reinforced moldboard bracing.

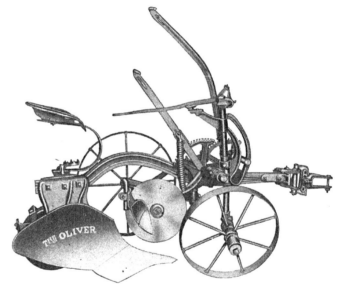

This dual plow design was an early attempt at a tool to make raised beds.

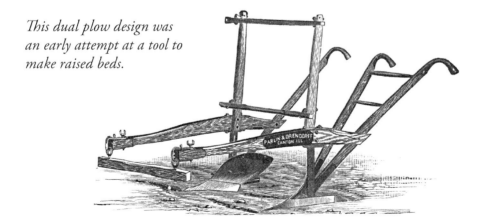

The next generation of riding plows featured a frame-style construction which had the plow hanging 'neath the three wheeled frame.

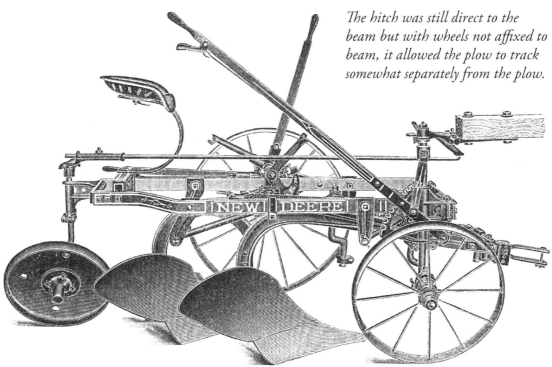

The hitch was still direct to the beam but with wheels not affixed to beam, it allowed the plow to track somewhat separately from the plow.

Oliver made one of the more popular two way plows. This implement allowed that the farmer could interchange beams at headlands so that he was always plowing in the same direction. This did away with dead furrows and made it possible to plow on the contour.

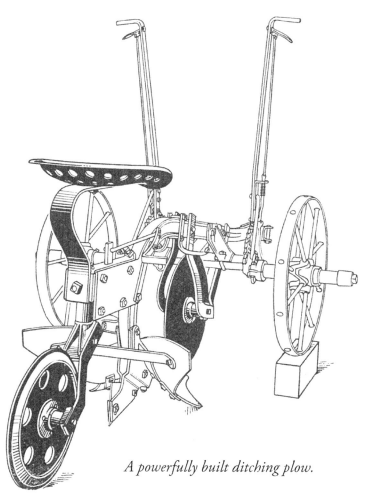

A powerfully built ditching plow.

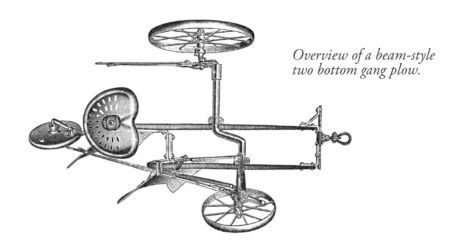

Overview of a beam-style two bottom gang plow.

The infamous John Deere Gilpin plow.

A slick walking plow featuring a landside wheel, thought to reduce friction.

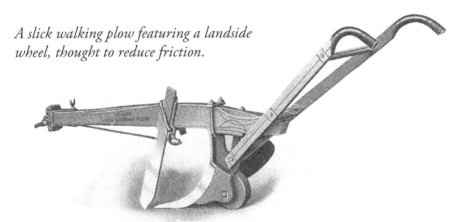

A road plow

Extra-heavy duty ditching plow

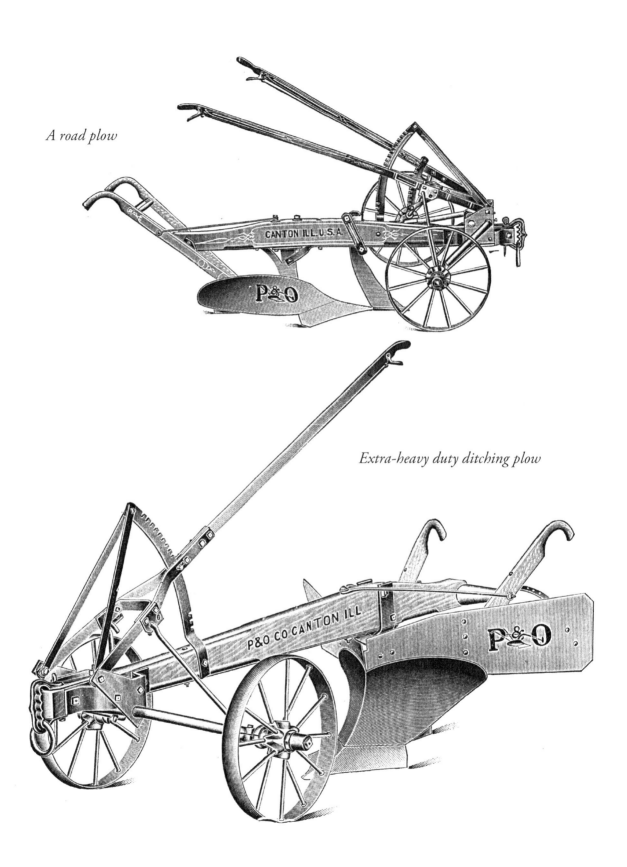

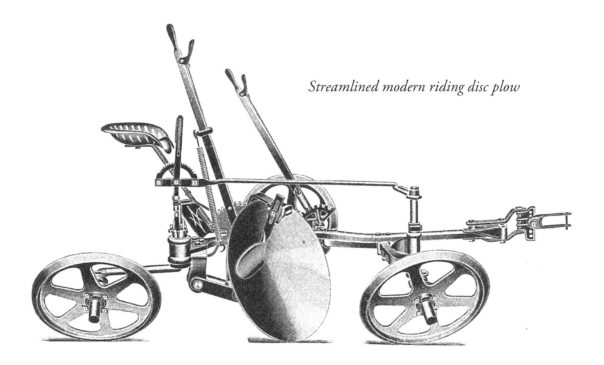

Streamlined modern riding disc plow

An overview of a reversible disc plow. When you get to the furrow's end you trip the lever and side-pass the horses around which turns the bottom to the other direction. I had one of these, the mechanism worked but it did a poor job of plowing and was dangerous on the headlands.

White Horse Machine manufactures this forecart, with ground-drive hydraulic accumulator, and this hydraulic two bottom plow, with breakovers for rocks and roots. Attached to this is a furrow harrow. Photos on these pages by author.

Another new White Horse plow design, both demonstrated at Horse Progress Days. Design such as these are clear and apparent proof that the future of real horsepower is well set. New companies would not be out there manufacturing these goodies if there wasn't a market and demand.

A view of the shipping lot at Pioneer Equipment in Dalton, Ohio. Dozens of new plows waiting to go out to dealerships around the country.

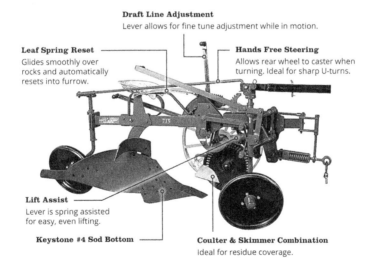

A features breakout of White Horse Implement's Leaf Spring Plow

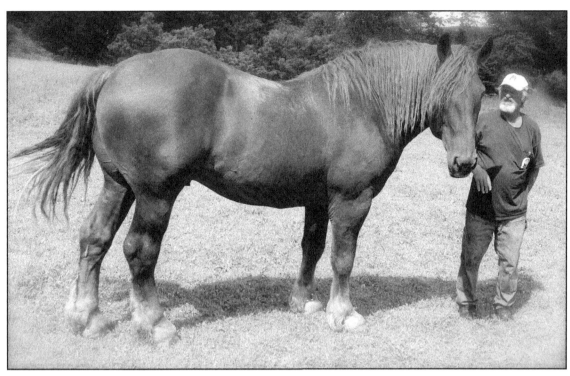

Horselogger and farmer Jason Rutledge with his Suffolk stallion,

Plowing old sod, Larry Bockus of Ontario, Canada

photo by Kristi Giman-Miller

Chapter Six   *"Will is to grace as the horse is to the rider." - Saint Augustine*

The Senses of the Horse

Levels of Communication

"If I've learned anything over the years, it's that horses do listen to you. They may not have a clue what you're saying, but they know the tone in which you say it. I'll sing to horses so hooked on their own nerves they're ready to climb into the sky, and sometimes it's one of the only things that keep them on the ground."
- Mara Dabrishus, Stay the Distance

Some people believe that all horses are dangerous, dumb animals. Stupid they say. They believe that these animals can only be trained through submission by restraint, by beating, tying up, chaining, and forcing into form. They are wrong. Horses are sublime, complex animals, of course quite different from us, but certainly not biologically inferior. If we are open to it, they can work with us even as well as a willing dance partner might. And a good dance requires give and take on both sides. You don't get a good dance with a pushy, arrogant bully taking the lead. And you don't get a good dance with a frightened, enslaved partner. For example, ballroom dancing is seen to have a leader - the male, and a follower - the female. But we know that much difficulty lies with the female's role, as she needs to understand or anticipate direction while working to have all the motion flow

Bulldog's six abreast

together. (Sometimes even with a clumsy oaf.) In a good dancing partnership the female may easily be the superior half without receiving due credit. Working with horses can be like this. We have the lead (or tell ourselves we do) and they follow. Yet given a chance to enjoy and contribute, the horse will add subtlety, intelligence, sensitivity, depth, range, beauty, and ease of motion (by anticipation and reaction) at fine levels we are often incapable of feeling or understanding. The above might sound like some description of dressage and hardly fitting

Seven abreast on JD gang disc, driven in PA. by a young Amish woman. Photo by William Castle

the needs and concerns of one who wants to plow with horses, but the point is that, athletic abilities differing, most horses share the sensitivity that allows for a calm, confident and wonderfully subtle working relationship. You don't need to yell at them, or hit them. You can whisper to them. You can touch them and they will move for you.

Jean Christophe Grossetette and Lynn Miller driving Belgian colts, Pat and Prince.

If you doubt this, someday watch a truly exceptional cutting horse perform, and measure for yourself the rider's contribution against the horse's. The best cutters realize the horse has a level of perception that is quicker and finer than their own. They must try to stay with the animal without getting in the way. The horse, if allowed, is busy reading the cow and positioning itself to prevent it from passing. If you have the opportunity, measure the above performance against one where the rider insists on making all decisions, and doesn't allow the horse to move without direction. It will become apparent to you that the horse has much to contribute. And we have much to learn.

For almost five decades I've been learning from my horses. I have more horse stories then I can ever hope to pass along. Some of them rise above the rest. One such occurrence was about thirty-nine or forty years ago. I had six head of horses, three and three, hooked to a two-bottom Oliver plow and I was out plowing about a ten-acre patch which I was going to plant to vegetables. In the lead I had a big, young, Belgian mare, Bobbie, in the furrow. She

Gina Werts of Tiller's Farmer Incubation Program leads Belgian gelding Prince. photo by Donna McClurkan

was an exceptional horse, beautiful if you didn't pay attention to her broken nose and her sad eyes. A red-sorrel, registered Farceur-bred mare I'd brought back from the Midwest. Bobbie and I got along real good. She took a little bit of patience because sometimes she would get confused. Plus the more patience I showed her the more trust she brought back to me.

Thinking back now, over all these years, perhaps I wanted to believe we had some sort of telepathic connection. Her actions and reactions reinforced that.

I'd been plowing for a couple of hours and had, on one end of that field, a narrow headland, I tried to maintain it as narrow as possible because I just wanted remaining an unplowed strip about wide enough to make a roadway along the vegetables I was going to plant. The opposite end of that plowed ground opened onto yet another field and there was no fence so I had plenty of room there to come around with the hitch. I was enjoying letting Bobbie help me on the narrow fence end of that plowed ground. We'd get down there to where we'd be just about to pull the plow out of the ground and I'd have to turn those leaders at the headland and still keep the plow going ahead straight; with the wheelers having to take the majority of the draft as I was swinging those leaders.

There wasn't enough room in that narrow headland for all six of those horses and the plow to come out of the ground so I had to fold the hitch. Bobbie made that job easy, she'd bring them on around. She wouldn't start to make her turn until I told her to. She'd turn "gee" or "haw" on command and she seemed to enjoy being in that furrow. We made this particular pass and I noticed that there was a white Cadillac that had pulled out into the field and parked off at the corner. There were two old gentlemen there watching me and I guess maybe I was showing off a bit, spoke to the horses a little louder than I needed to. We went on down and made that next round and got to the end. Normally I might have stopped there, where the horses had just pulled the plow out of the ground, and let them breath but

Tom Odegaard of North Dakota plowing with his six Belgians and a rope and pulley hitch. Photo by Fuller Sheldon

I swung them on around to get them into position and, as will sometimes happen when you are showing off, I didn't get the leaders quite where they belonged, Bobbie had come around a little bit too far from the line pressure and she was standing on plowed ground.

One of the men I recognized, his name was Ernie and he started talking to me, so I instantly just transferred my attention to the men and we talked for a little bit. Ernie, complimenting in an indirect way, was telling me some horse stories and I looked up at my horses and noticed Bobbie was standing on that plowed ground and not in the furrow where she belonged. Also I noticed that Bobbie's ears were turned and she was listening to Ernie talk. She was picking and choosing who she was listening too and there was a new voice out there. She was paying it a lot of attention.

Soon I was ignoring Ernie and, frustrated, I was thinking to myself, "Well, doggone it, I wish you were down in that furrow where you belong, Bobbie, because it doesn't look very good if, now that I'm showing off, there you are standing on that plowed ground instead of in the furrow" Ernie kept talking. She couldn't see me but I directed my attention at her, stared at the back of her head and thought, "Bobbie you need to be on over there in that furrow, you know that" and I watched her ears rotate. I hadn't said anything but her ears now were rotating around like she was looking for a different source, a different sound, and remembering the experiences I'd had with Bud and Dick (see last chapter) I just concen-

The author with three McIntosh Belgians on Sulky plow

Lynn Miller driving six McIntosh Belgians on gang plow.

trated on Bobbie. Ernie kept talking about this and that horseman and last Saturday's pull in Yakima. He went on and on. I was paying half attention or less and really concentrated on Bobbie and I thought, "step over into that furrow, girl." I was saying that to myself silently. "You know where to go, get down into that furrow."

Charles Orme and sons with big hitch of Belgians on Pioneer gang plow.

And her head came up just about two inches and she very slowly stepped over to the left and down into that furrow, just as clean as a whistle. Ernie noticed that I was looking at the mare and directed his gaze at her catching her as she was stepping into that furrow. He looked over at me and smiled and said, "that Mare, she likes to be in that furrow doesn't she?" What that mare liked, what she enjoyed for her entire long wonderful life, was being trusted. And the fact that we were both open to communicating with each other in different ways - kind of like a dance.

SENSORY
Keeping it simple: the horse's senses stack up this way.

Hearing: extremely acute and aided by ears which can rotate 300 degrees.
Touch: again very acute rendering some of them postiively ticklish as they can feel flies barely flitting on the surface of their hair coat.
Taste: it is suspected that horse's have a highly developed sense of taste sometimes making it difficult for them to drink water the taste of which they find objectionable or foreign.
Smell: we know that they can smell things which escape humans.
Sight: this is the oddest sensory realm of the equine. They have a dual focusing capacity with a slow restict retina and many fixed lenses around the eye that require for them to look out the top of the eye for close, the bottom of the eye for distance, and a rotating process to determine clarity in close to mid-range.

Suzanne Lupien with her team of Brabants

A scene at North East Animal Powered Field Days in Vermont. A perfect view of an intelligent horse moving straight ahead with the cultivator. Photo by the author.

A team of draft horses in France. Photo by Jean Christophe.

Horses are intelligent animals. They can measure an experience and make a decision for themselves. I remember when I moved to the coast farm, I moved from the valley and I took ten head of horses with me and it was surrounded by National Forest. There were different smells there, different things to see, hear, feel. The air smelled like the ocean. The horses were nervous for a little while. After a couple two to three weeks they seemed to be settling down some and I was out working by the barn and I noticed the horses snorting and looking off into the hillside. I couldn't see anything. There was a carpet of Douglass Fir trees there. Where they were looking was 3/8 to ½ a mile away. But they were snorting and wheeling around and they would run off about 10-15 feet and they would stop and stomp their feet and snort and run off. They did this for quite a while. There wasn't anything to see, maybe there was something to hear. That I couldn't make a point about because the human hearing comes nowhere close to how sensitive and acute the horses' hearing is. But I suspect that what was happening was that they were smelling something. Because about an hour later from the back window of the house the horses had all gathered together way off in the center of the pasture near a bog area and they were all looking in the same spot and when I looked over there to the fringe, an hour from when they had first been alerted, in walked into that pasture a small black bear, wandering around through the blackberry bushes and that black bear had quite an odor to him and I think the horses picked up on that odor.

About two years later I had an occasion for a bear to come into the pasture stumbling around and walking in circles. We found out later, the state trapper and I, that the bear had been blinded by a fever and he was very sick. But the horses wouldn't have necessarily known that

Jeroen Vos of France cultivating in his garden.

I don't think. Anyway, they had lived in that part of the country long enough to familiarize themselves with the smell and had already made a decision about the relative danger of a bear, and this bear didn't bother them, except for the fact that he was trespassing. And they herded that bear into that bog until he was standing out in the water and they pushed him. He couldn't see them because he was blind but they kicked at him and nudged him and herded him to get him out of their territory. They had acquired the courage to do that on their own. That first experience with a bear had terrified them but in there own good time and in their own good way they had measured that experience and made some decisions about it and demonstrated to my satisfaction their own intelligence.

The horses taste and their sense of smell are very, very acute, much more so than the human, and their sense of touch in incredible. You can watch how a horse's skin will react to the very smallest of insects touching it. The smallest insect just brushing along the hair coat at the leg or belly or back of a horse will cause that entire animal to twitch, maybe jerk, maybe swish its tail and arch its neck to try and push that insect away. They are so sensitive to touch. And the ears are amazing, the way their ears can rotate 300 degrees or better, picking up sounds from different places.

And the horse has a real appreciation for sound, and for rhythm. If you've ever had an opportunity to watch a dressage horse moving in perfect harmony with music then you know what I'm talking about.

Something about the nature of a horse which wants to match the pace of who they are with. Marie Brown leading stallion Modoc Herbert. Photo by Robert Mischka.

Comtois gelding. Photo by Jean Christophe Grossettete.

"One of the best pieces of advice I ever got was from a horse master. He told me to go slow to go fast. I think that applies to everything in life. We live as though there aren't enough hours in the day but if we do each thing calmly and carefully we will get it done quicker and with much less stress." - Viggo Mortensen

One way I choose to understand man's best potential relationship with horses is in terms of electrical current. Horses absorb electricity from everywhere and they give it up only when circumstance or teams connect. When I work my horses I feel a unexplainable difference in myself that I understand as hum without sound. I see this as the energy of the natural world coming to me through my working partners. When I truly believe in this connection, as I work, my relationship with the horses and our performance together is balanced and effective.

Shaun McInnis at the rear of a procession of plowing contestants at the Oregon Draft Horse Match in Yamhill. Photo by Kristi Gilman-Miller.

photo by Kristi Gilman-Miller

Chapter Seven

"Some people regard private enterprise as a predatory tiger to be shot. Others look on it as a cow they can milk. Not enough people see it as a healthy horse, pulling a sturdy wagon."
- Winston Churchill

"When I saw the movie, I said, I wish I had heard the music. I would have ridden the horse differently." -   *Eli Wallach*

Sharing the Load, Matching Time

It was summer time twenty years ago and I had been mowing for 4 ½ days with Cali and Lana, one of my favorite teams of Belgian mares. We were hitched to a number 9 high-gear mower, and had started out with about 50 acres to mow. That is a lot of mowing. It was early afternoon and we were just about done. We'd already been in for our lunch break and gone back out. We had about an hour's worth of mowing to do and a little difficult corner that had an odd shape to finish out, so there was a lot of back and forth. It went well, we finally finished that piece. It was one of those perfect summer days when the sky was crystal clear blue and there was a carpet of barn swallows following us around, harvesting the insects the mowing would chase up. Off to my left I could see the snow-capped Cascade mountain peaks. We were near the middle of our 160 acres of irrigated hay ground all surrounded by Juniper and Ponderosa pine. Coming out of those trees were some small Kestrel hawks and Red Tailed hawks working the field. About a half mile off I could see a Coyote chasing a

Amy Evers mowing with Red and Molly at Singing Horse Ranch in Oregon. Photo by Kristi Gilman-Miller

Sage Rat. It was a beautiful day, an 'all's-right-with-the-world' day and it felt powerful good to get that mowing done. We had about 5/8 of a mile trek across the mowed ground to go back to the barn and I had my legs crossed, heels casually resting on the frame, and the horses were walking at a nice even brisk walk. The walk they have when they know they've done a job and are going back. I had both lines in one hand, my other arm was resting on my knee, my hat was cocked back on my head, and I found myself singing. There was nobody to hear me sing and so it didn't matter if I was on key or off key, but I was singing. I noticed before long the rhythm of my singing was matching the pace of the horses and I imagined that I had done that subconsciously - you know, as in - they walked along and I sang a song that just matched that pace. But, I got to thinking, "maybe their pace is matching my singing". That seemed ludicrous. But I thought "well, I'll just try it," so I slowed down my singing, slowed it down about half the pace and the horses slowed down. Not only did they slow down but Lana laid her ears back turning her head and neck to look back at me as if to say she didn't approve of the new speed of my song. So I went back to the other pace I'd been singing at. She seemed to be happier and they sped up, matched the pace. I sang a little bit faster - and a little bit faster and they started to trot. I steadied them a bit, pulling back some on the lines, because I don't like letting them trot back to the barn. They start to get that idea and it just ends up being a hassle. Better horses are always trying to outsmart you and if

William Castle of Shropshire, England, mowing with a single Percheron mare.

you give them a chance like that they take advantage of it. So I slowed them back down and brought the pace to match the song I was singing.

It sure felt right then that all three of us were singing together. It sure felt that I was where I belonged, doing what I ought to be doing. I can't thank my horses enough for all those many moments they have given me. I felt complete and connected to the natural world.

I'm an old man now (protected in my seniority), but as I read these words back to myself I realize that, if you are young and just getting started, you might want to keep some of these types of self-revelatory experiences to yourself. Some people won't believe you and may even think you're daft.

MOWING

Horsedrawn mowers are clickety-clackety, whirring, shivering, rattling, ratcheting, squeak and pop machines. Horses pulling them for the first time can be forgiven if they get the

Mowing at Maple Hill Farm

jitters. If you haven't sat on one before while its working, you yourself will likely get the jitters. It's complex and it can make a novice pay attention to the wrong things in a tight time frame. If you don't know what you are doing, this is the bad machine to learn on. And if your horses are new to harness, most definitely don't hook up immediately to a mower. That said, mowing is one of those jobs where good work horses or mules, in a positive relationship with the teamster, shine. And it is a working rythmn and realm which rewards the attentive, appreciative farmer moment after moment. You can see your work up close and easily appreciate how a good team, pulling a well-oiled and timed mower, can make a complex and important farm job pure pleasure. Watching the hay crop cascade back as it is cut, seeing your self actually getting the job done, and at this comfortable pace, adds to the joy of the work. This picture is of the goal, or should be. And you don't get to this point unless you put in the time with your work animals, time and a teaching routine. Not just with mowing, it applies to all the working systems with draft animals - each and every single step you take, allow, insist on, all those increments must add up to an optimal relationship and efficient willing teams. The goal is made, not stumbled upon. You make it happen. That's the secret. And the patience and imagination and appreciation you bring to that process - they give you

Ken Geiss mowing with his little guys.

the art. Where people of the modern age frequently fall short is when they exchange accomodation for insistance. You let your horses have their way and trust falls apart.

It is all so very basic but MUST be repeated, and even those of us who have been at it a long while need to remind ourselves, every single moment you are working the horses is either a training moment or an unwinding. I make it a habit, whenever driving my teams into the barn, or through a gate that will regularly need to be opened and or closed, to stop (even if it is inconvenient) and make the horses wait until they have accepted and show no eagerness to contradict my wishes. Such a simple routine has paid me huge dividends in the long run. When I am mowiing a big field, the routine process of making a corner, for an intelligent animal or team, will be an invitation to act on anticipation. They see the corner coming and they say to themselves, "I know what's coming I turn right here" and so they set themselves to do that, they may even feel like that is what you want them to do. Besides messing with the cut, allowing the horses to do this on their own hands them control. Make it obvious that it is your decision to turn here, your decision that they can go now, your decision that they can stop here. Enjoy when it is obvious that they are waiting for you to tell them what's next. Trust that you may come to know your horses.

I & J Manufacuring of Pennsylvania has come up with some state of the art innovations in animal drawn mowers. Jason Rutledge and his Suffolks are using a ground drive forecart to operate a PTO trail mower.

A team of Bretons on an I & J Manufacturing enclosed gear mower with an eight foot cutter bar. The German company ESM, who make the double reciprocating cutter bars with the irreguarly spaced sections also had a stall at Pferdestark. Although they did not design the cutter bar for use with horses, being engineers they certainly appreciate its benefits for horses, due to the lower power requirements, and see a steady increase in demand for their product. Photo by William Castle

The author test drives a freshly rebuilt No. 9 McCormick mower with Renee Russell's Percheron team following a mower rebuild clinic conducted on Kenny and Renee's Mississippi farm. Photo by Renee Russell.

Old Ad showing factory colors and applauding "more than average performance".

There are still a great many very early model mowers stuck away in barns, museums and fence rows. They can be an enormous challenge to rebuild as the parts are difficult to find and/or replicate. This mower model was originally offered in the late 1800's.

A push-type lawn mower has been customized to take a swivel mount set of shafts so that a pony can be employed to pull it.

Eric Nordell utilized a manufactured set of steel shafts to convert this No. 9 to single use. Note the modern seat addition.

You can pick how far apart you have your teams working and on a hot day the little extra distance will lend cooling comfort.

Thomas Roberts of Lenoir, NC, mowing with 3 abreast. Photo by Paul Roberts Factory-built, three-abreast hitches suitable for mowers are few and far between, innovative farmers have designed their own to advantage. If the mower is properly timed, lubed, sharpened and ready a two horse team will handle it fine. If you have small animals, long hot days, steep inclines, or a poorly prepared mower, you may find use for a three abreast set-up

The rule with a three abreast is that you always put the quiet one in the middle.

Erik Andrus of VT built this bakery wagon to ply the streets of Vergennes with baked goods.

Chapter Eight

Sight and Rosie's Monster

What makes the working horse tick? We talked earlier about taste and smell, touch and hearing and how incredibly acute they are in the equine. How it is that the horse can hear things from much further away than we can, can feel the minuscule touch of a fly on their hide, how they can taste and smell things that are invisible to us. The eyesight however functions with complex inadequacies and shortcomings. The horse has a curious structure to the eye that causes limitations and challenges for them. I attempt here to describe that for you and I take some liberty with my description, I hope my veterinarian friends will excuse me for this. Imagine that the horse's eye has a cluster of little fixed lenses of various diameters organized all across the concave surface. In the center of that eye there is

Some years ago I had the distinct honor to assist with announcing at Horse Progress Days in Ohio. An Amish friend, who was driving twelve head on a four bottom plow for demonstration purposes, asked if I wanted to ride along. From up next to him, I took this photo. He and his hitch had to navigate a sea of thousands of people to get to the field and then to actually plow. He, as the teamster, could not see folks of all sizes and ages who might have been right in front of the leaders or just off the side a foot or two. He had picked his first choice of a lead four abreast, animals that would be fearless, move ahead on first command, watch their steps, stop when asked and turn on voice commands. He picked four horses he "knew". All of these horses had done thousands of hours of field work, and had differing amounts of exposure to crowds of people. On this day, and in this circumstance, the horses did incredibly well. Far and away better behaved than some of the goggled-eyed people who were oblivious to any potential hazard. I may have gotten myself in some hot water by observing to organizers that, even taking the obvious skill and mastery of the horsemen and the superb animals, unnecesary risks such as this environment represents can and should be avoided. I applaud that steps have been taken in subsequent years to keep crowds of folks back away from the working horses and implements.

one lens that functions very much like our very own iris, focusing in and out on those things that it chooses to look at. But the muscle is slow to react. It takes quite awhile to focus. Around that central lens there are lots of the other lenses, each one a little different. On each horse they are positioned in such a way that the animal can move its head, side to side and up and down, and get a fairly instant focus on something by selecting one of those lenses.

An Amishman ground drives his team of mule and gelding up a paved incline. Mule is casually looking to the side, teamster is checking left for oncoming traffic on a cross-street, and the gelding, head up, is trying to get a bead on something ahead and off in the distance. Three sets of eyes all paying attention to different things.

This explains how, when you see a horse out in a pasture and it appears to be looking way off in the distance, it will hold its nose way up high and look through the bottom of the eye. Or when a horse approaches a fluttering piece of paper or a puddle in the road, it'll lower its nose way down and appears to be looking through the top of the eye. The reason is that it's selecting the lens that will work. With this selection process it is testing lots of lenses. And, it is most useful to understand that in that process that it is getting some multiple images. While it's trying to find the lens through which it can focus on a puddle that is directly in front of it, the puddle may look bigger or smaller than it is. It may look deeper. It may even look like two or three puddles, until the horse finally rests a lens that will give it an immediate and accurate focus. And then, as the central lens of the eye does its job, the horse will bring its nose up and focus on using the center of the eye as it gains some assurance that it has finally figured out what it is looking at.

This dynamic explains the difficulties folks have with how their horses handle surprise, new stimulus and rapid shifts in environment. The horse tends to shy from something it sees immediately in front of it and the inexperienced horseman or horsewoman might be forgiven when it tries to force the horse through that problem, or take it away from that problem immediately, refusing the horse an opportunity to make a judgment of its own. Every time this happens you are robbing the horse, in that situation, of its opportunity to earn a little bit of courage out of that moment. Courage is earned when the horse figures out on its own that this is not a boogey man, this is not a deep dark hole, it's something that it can deal with.

The author with three McIntosh Belgians on a JD riding plow, in the Dufur parade. Steel wheels grinding on the pavement, lots of visual and audio distractions with parade entries, people along the sidewalk, children running out to get tossed candies, and the horses quiet and steady as can be. Photo by Kristi Gilman-Miller.

Picture a warm summer day on the coast range of Oregon, paved two-lane road, a beautiful black mare hooked up to a buggy, two of us out for a ride. Rosie, the Morgan x Arab, was harnessed and hitched in the shafts. Gosh what a fine mare. I used her to cut cattle and rope off of. Whenever I want to go out for a buggy ride or a sleigh ride, she's the one that goes into harness. She is broke for traffic like nothing I've ever had before. Loaded log trucks could approach us at 50 mph in the opposing lane and go zipping on by and she didn't care. She'd watch them as they went by, never shy. We'd have a loaded gravel truck approach from behind, give a toot of the horn and go on around and didn't bother her a bit. Dust could fly up at us, if we happen to be down a dirt road, that didn't bother us at all. Ah, Rosie, she's a good one; smart horse, courageous horse, top of the pecking order out in the pasture. She's become a legend for us. She's no longer around, passed on at 35 years old, but back all those years ago... I'm remembering an occurrence where she taught me a great deal about this whole horse's eyesight business.

See it with me, I'm going down that paved road, no shoulder to the road, either shoulder is a deep ditch and growing out of the ditch on either side, 10 foot high or better from above the pavement, are blackberry bushes thick with ripe berries. We come around that corner

On the farm, familiar surroundings, comfort and safety all around, horses will excel easily. But take them to town, to fairs, to parades, to field day events, and they are subjected to strange and ever changing sights, sounds and smells. Most trained work horses do fine away from home, a few can't handle it. (Same thing can be said of people.) Here, the superb McIntosh geldings are tied to the side of their trailer and waiting to be hitched to the header at the Dufur Threshing Bee. Photo by Kristi Gilman-Miller.

James McIntosh drives his sister, Janelle, to her wedding in the family cut-under, extension-top, surrey with Belgian gelding in the shafts. Photo by Kristi Gilman-Miller.

and as we hit a straightaway of maybe 150 yards, on the other end coming right towards us, a motorcycle. The biker sees us, buggy and horse and he figures "well, maybe I'll just wait and let them go on by, don't want to create any hassle here". So he stops his motorcycle and sets his feet out to hold it up, but he keeps the motor going, and he's got his headlight on. As he plays with that throttle to keep the engine running, pulsing back and forth, that headlight gets brighter and dimmer, brighter and dimmer. It pulses in perfect unison with the throttle. We're 150 yards away, Wayne and I know what's going on over there and I'm watching Rosie. I notice she's kind of ducking her head back and forth, back and forth, up and down and I recognize that she's trying to get a fix on that motorcycle. So I stop her for a second. Let her see if she can figure it out. She's moving her head all around and she seems like she's close to maybe getting a fix on it, but its still bothering her more than usual. And I can see by the movement of her head and the twitching of her ears that what she is reacting to is that synchronized pulse of the throttle and the headlight going back and forth, back and forth.

I know from experience that when Rosie has something thrown at her she'll take it on. I figured, "well just bring that guy on by." So I kind of stand up there in the buggy a little bit and flag him to come on by. He sees me and he holds his hand up and starts to walk his motorcycle forward a little bit, hit the throttle a little bit. He's still pulsing it, and as that motorcycle starts to come forward Rosie looks at it and she starts ducking, a full body duck right and left, like she's trying to figure out which way to go. Well, I'm steadying her with the lines and I'm talking to her, but the more I do that the less effect it seems to have. She's just not sure what the heck is coming at her. It's like a one-eyed dragon that is approaching her, and that pulsing light, synchronized with that throttle is just driving her nuts. Well, the guy riding the motorcycle seems to pick up on that and he stops again. And when he stops he plays with the throttle even more, to try and keep that cycle going. And that pulsing advances. Rosie starts to back up and I speak to her. I decide, well we're just going to drive on by if he won't. I talk to her and get her going forward, and as she's moving forward she's still paying real close attention to 'the monster'. Before I even know what happens she ducks off to the left and jackknives that buggy. When a buggy turns off sharp to the left or right the wheel can come up under your reach and tip the body. We almost had that buggy go over on us. Wayne jumped out and braced the buggy. And Rosie lurched right into the blackberry bushes on the road side, and tangled up there. Smart horse that she was she stopped fighting and stood still, waiting to be extricated. The motorcycle came up alongside real quiet and the guy was very apologetic and he went on by and I told him, "forget about it". We pulled Rosie out of the blackberry bushes, she was none the worse for wear but a little bit scratched up and nervous. We got her all lined out, but it taught me a real important lesson about how, even with the best of horses, the eyesight can work against them at times. How you need

Photo shared by Dick Brown.

to understand that and use your intelligence and try to give that horse the best opportunity possible to measure and know the threat they see. But we need to understand that the perfect horse can get thrown into a tizzy when all the circumstances prove disabling, and how they see what is worrying them may be the first place to go for understanding.

With time and repetition horses come to learn their working environment AND how much they can depend on their teamster partners. If you don't get easily rattled, they pick up on that and it gives them strong reassurance. If you, on the other hand, get fidgety and loud or quavering and uncertain in your movements they definitely pick up on that as if to say "can't rely on this bloke, need to protect myself." If you go overboard and coddle your horse, coo at them about the hazard, plead with them and then take charge, you think, by removing them from the hazard altogether, you've made matters worse. If they see you walk, sure and certain right through that puddle, they will be far more inclined to follow on the lead. If instead you stop at the puddle and say "hey big boy, it's just a puddle, nothing to be frightened of. Here see? You can do it. Come on now." When you get sappy with your accomodations you've allowed it to become a battle you can't win. Take charge invisibly but with certainty.

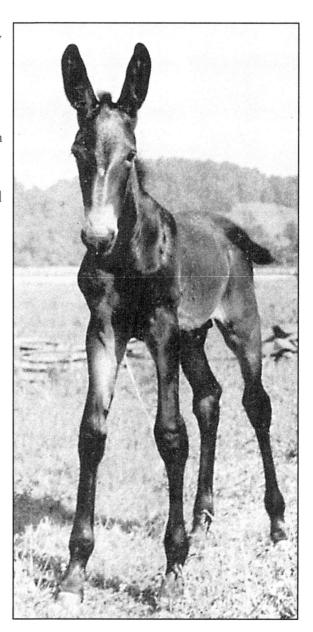

In all these things for me it keeps coming back to this one distinction; I don't want to know that I can trust my horses, I want to trust that I know my horses.

"The mule is jus like a horse, only it ain't"
  - Bowell Adroit

Four abreast of Belgians ground-driven back to barn at 2016 Horse Progress Days, photo by Jerry Hunter.

The Peterson four-up of Belgians at the Dufur Threshing Bee parade. Photo by Kristi Gilman-Miller

Photo by Bernard Chamber of Jonathon Waterer with Shires on hay wagon, North Devon, England

Photo by Richard B. Hicks. Mike Denny from Sauk Centre, MN with his Percherons Chip and Steel on McCormick Culti Packer.

Chapter Nine

Harness Essentials and Strip Pulling

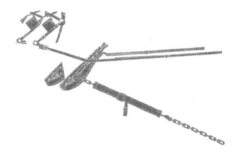

Imagine a horse standing there, a web of leather straps, some pieces of metal, some pieces of chain all draped over in a design; a design that seems to be geared towards a certain function. It is. The different parts of that harness are there for a reason. It's important if you're going to be successful and efficient at working horses to understand all the mechanics of the harness design. You've got two metal or wood ribs seated in grooves on the collar about 2/3 of the way down (the hames), there's a tug or trace fastened in, it hinges there and rides over the wide part of the collar, goes on back at a particular angle and transfers into a chain that hooks to the singletree. That singletree might hook to a doubletree or tripletree or whatever other evener you are using to get your job done. That tug or trace, as it passes alongside the horse is held at a particular angle by its interface with the belly band billets, and the back pad billets, depending on whether or not the harness is going to be used to pull a wagon, or something else with a tongue in it, you might have a backing or braking mechanism involving a brichen and quarter straps, a pole strap and a breast strap assembly. If you are logging or just dragging something along the ground you might not need that. But if you are going to be successful at working with horses you need to understand all of those parts and how they function.

In the great Pacific Northwest we have our share of legends in the work horse realm. I've had the most excellent experiences working alongside some of these gentlemen. And one of the funniest teachers and mentors I've ever experienced was a man named Dan Kintz. Dan Kintz was a horselogger extraordinaire. And back in the seventies when I was still involved in pulling matches I attended some contests with Dan. He taught me quite a bit. In a humorous way he gave me a technical grounding. We were at the Polk County Fair in Rickreall, Oregon. It was a summer evening. I know it was dark and there were about 400-500 people in the bleachers who were real nuts about pulling horses. I think there were 6-7 teams there that evening and we just had a middle weight and a heavy weight class. I was in the middle weight class with Dan and most of us were real good friends. We did it as a sport. We liked to show off what these horses could do. And Dan, well, I think maybe, just maybe he might have had a beer too many that night. He got an idea in his head of a special kind of a side challenge at this pulling match. He challenged the rest of us to join him and we didn't quite have the gumption, because we knew on a dare he would take it all the way. There was something that happened at that pulling match that taught me a whole lot about harness function and in a most interesting way. Dan stood up front and announced over the microphone, to the crowd and to the rest of us teamsters a challenge, he called for a strip pull. He said, "I'll tell you what we'll do. Every time my horses make a successful pull I'll take one piece of harness off and I'll remove one piece of my clothing and I'll still beat you all." The crowd went nuts, but the rest of us really didn't want to get involved in that because as I said, Dan had

Legendary horse-puller, Harry Faucett.

Jason Rutledge and his pulling Suffolks. Photo by Suzie Kelly

a reputation for going all the way. In fact it wasn't too long before that; on a dare Dan had taken one of his draft horses in harness, Streak, a Percheron gelding, right into a bar in the town of Mt. Angel. And he proceeded to order a beer for him, did it on a dare. Bartender said, "you're nuts, take that horse out a here." And Dan said, "I'll make you a wager, if my horse will drink a whole beer then I get beer tonite for free. If he don't, I'll buy the whole house a round." Dan won that bet.

So, now we're back at that pulling match and Dan has laid down his challenge. We went ahead on that middle weight class. We were at about 750 lbs. just to get started with the pulling match and sure enough Dan's horses walked away with it easy. He walked over to the harness and proceeded to take off a quarter strap and then he pulled his shirt off. He had a t-shirt on underneath it. And on the next pull we did 1,000 lbs. and he took off another quarter strap and just for good measure he pulled off a pole strap and he took off a shoe. And he went that way on each pull until he had that harness down to where there was just the hames, the tugs and traces, those hames set in on the collar, and he had a back pad on and he had the bellyband. All the rest of the harness had been taken off and Dan was down to his t-shirt and his undershorts. There was only one man out there who was still in the pulling match and they were up to a heavy load. It was up to about 3,500 lbs. on that stone boat. The rest of us had already bowed out. Our horses had had enough. Al, fully clothed and refusing the stirp challenge, had the next load to pull and his horses couldn't do it. Then Dan was up with his team needing to strip another pair of items.

Nobody could imagine that he'd be able to pull the load if he took any more parts of the harness off. I was curious what he was going to remove. The crowd was entertained and anxious to see what other piece of clothing he was going to take off. Here's this man, he's got a red neck from working in the sun and he's white all the way down and he's wearing a pair of boxer shorts and a t-shirt and the crowd is just loving every minute of it. And he walks over to his team and he proceeds to take a back pad off of one horse which meant he only had the belly band attached to those tugs. He took the back pad off the other horse and he pulled that t-shirt off. With just his boxer trunks on and that harness stripped down to nothing more than collars, hames, tugs and belly band, he pulled that winning load. The crowd went nuts.

I talked to him later and I asked, "Dan, why did you choose to take the back pad off?" He said, "Well you know how those horses are they get into a heavy load they get right down onto the ground and squat and scratch and I knew I had to keep that tug angle right, because if those tugs were raising up too high it would choke them and they wouldn't be able to do the pull."

Tongs set on the log. Photo by Kristi Gilman-Miller.

Chapter Ten

Logging and Woods Work

Rube the Teacher

I knew a man who went into the woods with a big gelding "Rube", some hand tools and very little experience. He was joining a two man crew, one with a single horse and another with a team, to skid logs to a landing. When he applied for the work they asked him if he had experience. These horseloggers were too harried and busy to listen to his answer. "Not in the woods," he said. "Ok, just follow our lead."

Given a gentle downhill slope, a fresh snowfall, and a good horse, big logs may be moved from woods to landing. The woodlot on Natural Roots Farm. MA.

At an Indiana Horse Progress Days a few years back, Jim Brown and his Belgian team demonstrated the functioning range of a custom-built logging arch; a tool which allows the front end of a heavy log to be lifted enough to dramatically reduce the draft friction. Photo by Kristi Gilman-Miller.

For that first day he did just that, stumbling and fidgeting, fighting his nerves and lack of knowledge. He got in behind the other single horse and hooked to the smallest logs he could find. When the other man and horse set out, he was sure to follow not too far behind. They took the logs to a landing and unhooked and went back up to the cut to get another. His horse did fine, perhaps because he knew the work, but the young man couldn't tell or was too wrapped up in his own experience to notice.

At one point the novice, on a new skid road, managed to jam the log in a triangulation of trees. When the log wouldn't move he got upset and yelled at the horse, slapping him with the lines. The horse jack-knifed and straddled the log heading in the other direction. "Stop right there!" hollered the other logger, who proceeded to unscramble everything. "Holy monkey, man! You could've had a terrible wreck there. Do you have a clue what you are doing?" He proceeded to rehitch the log with an encircled bite of the chain and slip hook, then he spoke to the horse, calming him down, and pulled a few inches sideways so the log would roll and free itself. Stopping the gelding, he handed the lines back to the novice. "Pull yourself together man!" And he walked back to his waiting horse.

A big jag of two hardwood logs is pulled handily by two horses on a logging arch. The snow is the secret positive, allowing a slippery surface. So long as steps are taken to protect the horses from slipping while exerting themselves (common sense, patience, appropriate shoeing) snow is the horseloggers ally.

Below: Les Barden with homemade forecart.

The last thirty years have seen many shade tree innovations in horselogging arches. Topographical variations of local woods, tree species, and saw-log market demands all have had bearing. You don't need a high arch if all your material averages 8" diameter, and a low narrow arch doesn't do much good with 30"+ diameters.

The young man set the lines down on the log and wiped his sweaty face at exactly the moment that the other horselogger spoke to his horse to go. Rube was sure the command was for him as well, so he went ahead and the young man stumbled and fell reaching for the lines. Horselogger number two stopped his horse and went after Rube's lines as he walked off with the log. The second horse, not wanting to be left behind, started up and followed in behind Rube, both now moving at a brisk walk. The skid road meandered through the standing trees, so the two men went cross-country to catch their horses, the younger man in the lead, leaping wildly and hollering in a nervous, quavering, screeching voice "stop!"

The third horselogger, boss-man with the team, was at the landing and could see the two loose horses coming his way. He had the presence of mind to unhook his team and drive them around and tie them to the backside of the waiting truck. He then went around to see what was going to happen.

Rube slowed as he made the final approach to the landing, and stopped in exactly the right spot with the log. He stood still as the second horse came up behind and stopped.
The boss-man, the one with the team, couldn't decide whether to laugh or be angry.

Dick Brown and friend in procession through snowy woods with two logging carts. Photo by Robert Mischka.

Two pictures of Elwin Wines, SW Oregon logger. Top: With Suffolks Mik and Cedar yarding a heavy, 32 foot Doug Fir, down to the landing. Below: Fritz, the Belgian, joins the Suffolk team to skid good-sized Fir.

All of them standing together the boss said. "Listen to me carefully. You," and he pointed at the young man's chest, "are fired. Are you listening to me? Look at me. You are fired, but I want to rent your horse from you. And I'll allow you to stay here and cleanup stuff and watch, but you cannot touch your horse so long as he works for me. Got that? I suggest you take my offer. It's just possible that, if you pay attention, you might learn some things. Oh, and by the way, think about how much money you might take for Rube. After this job is done I might want to buy him."

The men then set up a routine where Rube, trace chains hung up and single tree on hame, would follow the one single horse outfit back up to the woods. He would hitch his horse first to a log, then Rube. He'd speak to Rube who'd proceed back to the landing sans teamster but with the log. The logger, his horse and log would then follow. They got so good at this routine that the two men would sometimes split up, one man on the landing, one man in the woods; one would hitch the logs and the other would unhitch and send the horses back up.

The young man sold the gelding to them and gave up on logging. The other two made a go of it for many years. When one of them told me the story I said, "My, my". I shelved that story until now. You can take from it what you will. I don't expect you to believe it.

Elwin Wines drives Ace and Deuce with three logs in the beautiful Applegate country of SW Oregon.

Here are two old-style, heavy-duty, bobsled building plans of the sort you might find in New England and the Maritime Provinces of Canada. These are designed to haul cord wood on the sled frame. Loggers and woodcutters from the north woods have long known snow and ice to be a friend to heavy skidding, as the old photo on page 200 will attest. The sled under this massive (and dangerous) load is of the same basic design as the two plans shown below, except for the fact that it is designed, with curved log bunks, to received full logs lengthwise .

Notice in the plans (page 203 for eight food wood, page 202 for four foot wood) that these sled set-ups employ two autonomous "bobs" which are connected with a length of suitable chain running from the front of each rear runner, criss-srossing, on to the main-frame ends of the front bob. This allows for a simple articulation on curves so that the bobs "round" the corners instead of dragging across. When the front bob turns left, the cross-chain tightens and turns the back bob slightly right (in a following action). This bobsled design is particularly elegant in that it features, on both front and back bobs, a swiveling cross yoke. On the front this means the tongue swivels up and down easily over rough terrain. On the back, the cross yoke ends have axle pins which fasten by plate to the ends of the reach chain which allows that there is no binding when running up and down over cross ridges in the skid. With the design on page 203, the frame is anchored by two heavy bolsters which each pin into the sled frames. These bolsters allow the bob to pivot 'neath them in the same fashion as a traditional wagon running gear. This sled will receive eight foot logs either lengthwise, in two bunk loads, or crosswise for the entire length. The runners are lined on the bottom with steel.

The plan on the next page (202), for four foot, cross-ways, wood, is designed for a single horse and narrow skid trails. It features shafts instead of tongue and has a lower profile. In this variation the front bob is affixed to the rack and the back bob, cross-chained, is allowed to pivot under the frame on turns. The same "following" aspect occurs here.

Singing Horse Ranch feed sled atop bobsled running gear. Photo by Kristi Gilman-Miller

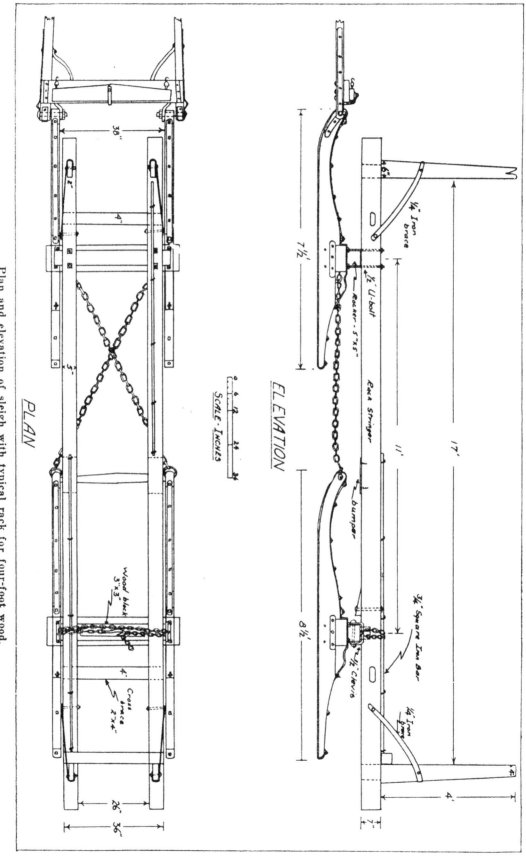

Plan and elevation of sleigh with typical rack for four-foot wood.

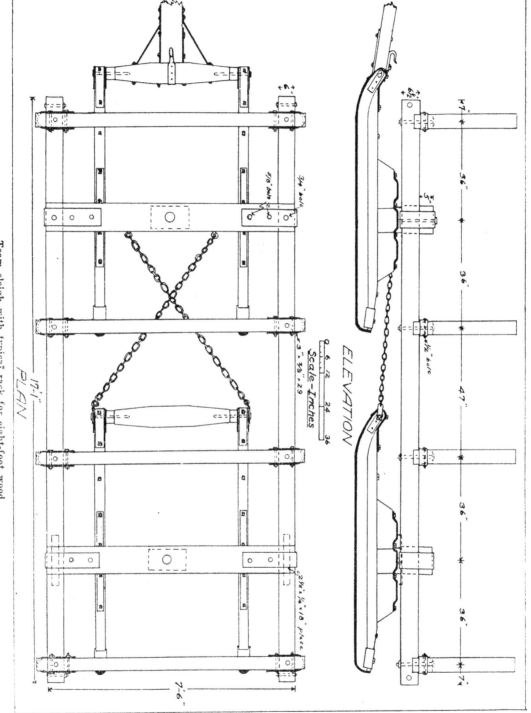

Team sleigh with typical rack for eight-foot wood.

Left: Jason Rutledge sets chain on an arch-lifted log. The dog is advising him. His Sufffolk horses stand patiently.

All aspects of woods work are extremely hazardous and require agility, ATTENTION, intelligence and patience. You need to always be scoping out where you are in respect to the landscape and the work. Uphill or upgrade from the log is a smarter place to stand. Avoid feet and legs in the bite of cables, or chains, and avoid standing with waiting roots and debris set to trip you up. Understand what the trajectory might be if and when the load moves ahead, even just a foot, without you planning it. It might be a horse stepping forward, or something giving way in the forest floor. Experienced horseloggers learn all sorts of tricks to navigate the big and little tight spots; one is to set a chain or choker so that forward motion causes a sideways turning action, this can help to start a heavy load and avoid a stump or rock.

Carl Russell and friend use peavies to roll a Vermont log up onto a log scoot. Photo by Lisa McRory.

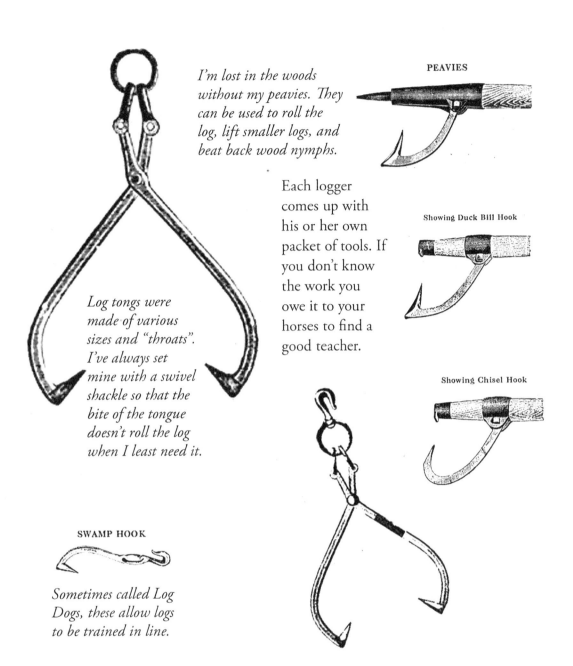

I'm lost in the woods without my peavies. They can be used to roll the log, lift smaller logs, and beat back wood nymphs.

PEAVIES

Showing Duck Bill Hook

Showing Chisel Hook

Each logger comes up with his or her own packet of tools. If you don't know the work you owe it to your horses to find a good teacher.

Log tongs were made of various sizes and "throats". I've always set mine with a swivel shackle so that the bite of the tongue doesn't roll the log when I least need it.

SWAMP HOOK

Sometimes called Log Dogs, these allow logs to be trained in line.

Log chains uusually feature a wide mouth slip hook at one end that allows the loop on the log to cinch with forward motion

COIL LOG CHAINS

Jay Bailey on one style of logging arch which provides lift as the horses move ahead.

Sam Smith demonstrates another style of arch which employs a balancing lever principle to the lift action. When the tugs are tightened, as shown, the device lifts with added force. When the horse backs up the log is lowered for detachment. This style of arch, which carries much of the log, moves the steerage back and translates to a different positioning when navigating through standing trees. It will take a bit of getting used to, expecially if you are accustomed to "feeling" where the drug log will travel.

A Canadian company designed and builds this complex horsedrawn self-loading log truck with its own onboard motor for the hydraulics. The cart front features auto-steering and out-riggers for stabilization as the hydra fork swings out to lift logs. Such an implement would allow that a team could move more logs a much longer distance and has been employed with success as an intermediate transport, from woods landing to highway truck landing.

Here Paul Schmit of Luxemborg and the non-profit Schaff mat Paerd, is testing a new Timber Wagon, the Osterby Smedja SV5 Forwarder built in Sweden. Paul, and Alban Moscardo of Italy, work to support the development of new animal drawn equipment.

Elwin Wines and team break out of the forest shade and into the open field's sunshine.

Chapter Eleven

Trust and the Ultimate Relationship

Lana's Bridle and Abe's Rebirth

Jessica Lindsay, young teamster of Oregon.

You can have the best harness in the world, brand new harness, all the parts fitting just the way they ought to fit, everything in the best of shape, and that's still no guarantee that you won't have a mishap. For me, the most important thing is to have a relationship with those horses, that's born of trust. You need to be able to trust them, just as much as they need to be able to trust you. But perhaps most important of all is that you need to trust that you know your horses.

I was out mowing hay with a mower that I had just reconditioned, first time out in the field. I used to restore No. 9 mowers. I had taken one completely apart and put it back together with new guards and wear plates and shims and bearings and all the rest of it. Hooked up my good team, Callie and Lana. They had mowed more acres than I could count, and we went out into a big beautiful crop of clover/alfalfa and grass. I was testing this mower to see if everything was working the way it ought to work. Moving along, I'm paying close attention to the mower; I don't need to pay too much attention to this championship team. I got both lines in one hand, the other on the lifting lever. Lana's walking right along the cut edge of the grass. Callie's walking along next to her. I've got the cutter bar down and the mowers engaged and I can hear the familiar clicking but I hear another noise in there: kind of an off noise, and I can't figure out what it is.

Intent, I'm looking down at the cutter bar, and back and forth at the mower, and trying to figure out what's loose, what's wrong. Something is not right. And I'm paying real close attention to this as we are moving ahead. The horses are just walking along nice and sweet, and then, of a sudden, they start walking faster. I don't bother to even look up at the horses, I've got a ¼ mile row to go and I can see from where I'm looking that they are cutting in the right place, so I just speak to them, "Callie, Lana, now, now, slow down a bit. Easy now girls. Walk. Walk." And they slow down and I keep watching the cutter bar and I'm trying to figure out what's loose. Do I have a guard loose somewhere? Is there a shim that's not right? What is it? Maybe there's a washer that needs to be put in on the axle. I can't figure out what that noise is. And I'm still watching and pretty soon those horses are starting to trot.

And I said, "Easy now girls, slow down!" I don't even look up at them. I've still got both lines in my hand. I'm not looking at the horses at all. I trust them completely. And they hear me and they slow down. And I'm watching that cutter bar again. I can't figure it out. What on earth is that? And then I see it. I could see that a guard kind of shifted off to the side a little bit. I've got a guard bolt that's loose. I thought I had them all tightened up. I can't figure that out, so I say, " Whoa, now girls.", and they stop and I bring my head up to look at the horses and put the mower out of gear, because I'm going to get off and get my wrench to fix that guard.

And as I look up, something doesn't look right. Sure enough, Lana's standing there, and she hasn't got a bridle on. It's hanging down around her knees! Holy smokes! How could that have happened? So, I take the mower out of gear, walk around, get around in front of her, and I take that bridle and I can see that something has come undone. I thought sure I had that throat latch on right but apparently it had come undone. Maybe I didn't hook it. I put the bridle back on her, got it all hooked up right, and then I stop and think about it. Many is the time I have driven uncertain horses where I knew I had to have everything about the harness right or I was going to have a wreck. I couldn't even think, before that moment, of driving horses on a mower, let alone on a wagon or anything with a pole, and having one of the bridles come off or break. They would have been gone, they would have just taken off. Not only did these horses not take off, but I don't know how long the bridle was off. For at least a ¼ mile I was driving those horses just by my voice commands; Lana without a bridle. I had come to trust them that much that it didn't even enter my mind to worry.

Matter-of-fact, I just walked up and fixed the bridle. I knew this team, I knew Lana. I knew what she was capable of. And that's what I mean about that trusting relationship. That's the kind of relationship you can and should have with these horses. But you MUST earn it. It takes time and an open mind. They'll give it to you but you have to ask them for it, and if you'll forgive a little cheekiness, you have to hold your mouth just so. You trust them, they'll trust you. And, the double reward is, they'll come to know you and you them.

Another time another story; Abe's Rebirth

This is not an easy tale to tell. Fact is I don't think I have ever written it down before.

Abe taught me what it takes to earn back the trust of a horse. And he also taught me what it means. There are those of us who want to believe that temperment is a highly heritable trait. But who are we to say we know the intrinisc, or core, temperment of any given horse?

There are horses who are so dull-witted as to be clumsy and a hazard to themselves, as well as others. And there are those who have a wide angry streak, discontented with everything and looking for a fight. And there are horses who simply want to be left alone, and will protect that with a fiercesome insistance. There are horses naturally thin-skinned, skittish and frightened of the least little thing. Another category are those animals who are smart and sneaky and perhaps bullheaded. And then there are intelligent sweet-tempered horses who want to please. All of these traits and more might be either essential to the horse's natural personality or a manifestation of conditioning and environment. The best horsemen can read the difference, though it may be of the subtlest distinction.

Abe, who I raised as a colt, was of the sweetest, most intelligent and willing personality - up until his second year. Then everything changed and he became frightened, mean, bull-headed, and prone to flight. It was all brought about by circumstances completely out of his control. But first a note of his breeding and background: he was the off spring of my outstanding Farceur-bred mare Bobbie, and my stud Melodist DuMarais (a colt I purchased from Jiggs Kinney and out of the legendary Eddie DuMarais). From the moment Abe was born, I had my arms all around him. He was a sweetheart, loved to be around people. Unafraid. I raised

Abe, charcoal and walnut ink drawing by Lynn Miller.

him that way.

As a yearling, I kept him in a boxstall with a ramp to his own outside paddock. Fed and handled him everyday. At that stage I had a guy working for me, didn't know him well. He was to take care of the livestock as Kristi and I had a long cross-country trip planned to cover events, do a workshop and purchase some items at midwestern auctions. We were gone more than two weeks.

When we returned we learned that Abe had disappeared. He had escaped and run into the dense coastal woods and nobody could find him. I went to the neighboring farms and asked for help to form a search party. My hired hand said he wouldn't help, 'cause he said "that colt has turned mean, he'll kill you, I say. If you find him you need to shoot him before he hurts somebody." That didn't make any sense.

In a short time, we found a neighbor who said the horse was at the backside of his farm, that he had tried to catch him but the horse came at him striking with his front feet. We went with help and found him, herded him onto the road and back to the barn where we penned him up. Not even I could get close to him to put a halter on. It broke my heart to see him this way. I wondered if something had gone wrong in his brain chemistry.

Then the horrid story started to unravel itself. The hired man had been meeting a teenaged girl in my barn at night. When she broke off the relationship with him he went nuts. While we were gone, he would return to the barn each night in hopes that she might change her mind and meet him again. Abe, in the barn and true to his nature would try to nussle the distraught man. The man got so mad he would take his belt off, and buckle exposed, would beat Abe's face til it was bloody. He did this for a week. Abe decided to defend himself and it all got much worse, resulting in the horse breaking the fence and running off to the woods.

Flash forward several months, bad man long gone, Abe is in his boxstall and paddock, still mean as a snake and my heart is still broken. I stay patient and try every long, slow trick I can think of to get him to accept me, but with no success.

We had a innocent young visitor at that time, Scott Borello. He wanted to learn about horses. He arrived that evening and I was in the barn milking the Jersey cow. I didn't hear him come in. Abe was in the corner boxstall, sulking. I heard a voice and looked up and saw Abe's stall gate was open! I trotted over and there was this guy standing in the stall next to Abe petting him. I had the presence of mind not to over-react. I said quietly, standing at the stall gate, "Hey, would you come here a minute?" Scott walked over, Abe following close behind. Soon as the boy came through the gate, I closed it slowly, thinking fast. I had an idea, "I need to

Joyce Sharp's Belgian cross team. Photo by Kristi Gilman-Miller

finish milking the cow, would you mind putting this halter on him? Unbuckle it and put it over his neck, don't pull it past his ears. And keep this gate shut."

"Sure." he said. And I slipped back to the cow, looking from a distance and keeping my mouth shut.

Scott walked into the stall, calm as you please, and proceeded to put the halter on. Abe seemed to be his old self. I returned, thanked him, and put some grain in the stud colt's feeder. Abe came over to it, ears back. I ignored him and walked over to Scott and visited with him. I did not give Scott any background information until later. And I noticed, when he went back to the barn with me in the morning, Scott was nervous and Abe was angry. They avoided each other. (After I had told Scott of Abe's problem he became apprehensive.)

A few days passed and I recalled that experience I had had years before watching Ray Drongesen with the angry stallion, Don Degas, and how Ray had been so matter of fact. The resulting long term transformation of that angry black stallion had been phenomenal. I was determined to follow that lead and try something. I went to the barn and gathered up several flakes of straw and went into the box stall, avoiding all eye contact with Abe.

I put my mind elsewhere, thought about my sawmill and a lumber order I had to complete,

Abe in the Sun, charcoal and walnut ink drawing by Lynn Miller.

and proceeded to scatter that straw on the floor of the 12 x 14 foot boxstall, ignoring that snorting colt. (I tricked myself; I was thinking about taking a team and skidding some Douglas Fir logs to the ramp on the backside of the sawmill, and how I had to repair my broken peavie handle. In my head I was elsewhere.)

The horse ran out the door and down the ramp. I got my pitchfork and proceeded to clean manure out, and he came running back up the ramp ears back. I forced myself to ignore him til he stood within a couple of stomping feet of me, again I was avoiding eye contact. I matter-of-factly reached up and took hold of the halter that had remained on him, and held his head still. He tried to yank away, but I put my chest against his face and held him there and he quieted. With the other hand, I leaned the fork against the stall wall and slowly stroked his neck until his ears relaxed. Then I slipped the halter off and let him go, picked up the fork and walked out of the stall.

When I turned to close the gate, he was standing right there, ears up, clear-eyed. Then he turned and nosed the fresh straw on the floor.

With no agenda, I passed each day, coming and going from his space. Occasionally I would pet him. A couple of weeks passed and I took a baling twine and slipped it up and over his neck and left it there for a few minutes. Next day I calmly haltered him. Two weeks later I led him from the stall into the barn alley and carefully, methodically, put a harness on him for the first time. He never moved a muscle. Never showed any discomfort or anxiety. It was as if he were born broke, born well-trained. As if he carried an incredible desire to be grateful, useful and appreciated. And he was. He grew to a magnificent stallion with his mother's dispositon, he was another "cantalope" of a horse - round, sweet and ready.

One of the reasons I have naturally been reluctant to share this story is that I fear people might take it as a suggestion for how they would approach a mean horse. It is most emphatically NOT that. I share the story to try to make a point. That point is if we can bring ourselves to trust that a horse's innate nature is good, just possibly we can help him or her work through the baggage that has stained that positive core.

But a horse will outweigh you and be capable of rapid, powerful movement that just might hurt you badly. Do not go there without experience and courage and a usable bit of stupidity. Or better yet, do not go there at all. Best, instead, if you find your capable self 'there' when it is necessary.

Suitable caution is one of the colors of a long life and deep experience, but also, so are richest memories of patience and sensitivity rewarded.

Abe on Singing Horse Ranch, photograph by Kristi Gilman-Miller.

"There's a way to train a horse where when you get done you've got the horse. On his own ground. A good horse will figure things out on his own. You can see what's in his heart. He won't do one thing while you're watching him and another when you aint. He's all of a piece. When you've got a horse to that place you cant hardly get him to do somethin he knows is wrong. He'll fight you over it. And if you mistreat him it just about kills him. A good horse has justice in his heart."
- Cormac McCarthy, Cities of the Plain

The good ones advertise themselves as such. Photo by William Castle

Chapter Twelve

Deep in the Mechanics

The Broken Strap

It's one thing to understand how an individual harness functions on a horse, to get it properly adjusted so the horse is comfortable and can work all day without any kind of sores or aggravation. It's another thing to understand how you might put two or more of those horses together in harness and hook them up to a doubletree or tripletree, or whatever apparatus you might be using, to pull whatever implement it is that you're going across the field with. And then it's quite another to understand the geometry of all that, with a tongue attached to a backing and braking mechanism, and maybe you working a grade of some sort where the draft is difficult, or pushing against the animals should you be going downhill. You have to understand these dynamics. There are times when your safety will depend on it;

Dick Brown of Iowa, mowing with his Percherons. Photo by Robert Mischka.

all that geometry, how it goes together, functions and what you need to do when something goes wrong.

Red's one of my favorite horses and that will be true until the day I die; half Belgian, half Percheron mare, a Bay Roan. Page 17 of this book features a pastel I did of her. I pictured her on the cover of one of my books. She was unusual, one of those mares that just doesn't like to be treated as a pet. She wants you to do what you've got to do, and get it over with. She just wants you to leave her alone, but by gollie is she a worker. She'll pull all day long. She can get nervous, but she's one of those horses, like Bobbie I mentioned elsewhere, that will give you big dividends if you trust her. If you are patient, she'll return it to you. Talking before about that business of the geometry of the hitch and how important that was to understand; I wanted you to know a little about the mare that features in this story.

Mollie, Red and I were mowing the upper hay field and there's a grade there, not much, but a bit of a grade. (You might be surprised to find out that one of those horse drawn mowers can weigh 600 - 700 lbs.) This mower was working great. We were cleaning up a kind of headland area, right in front of where I was going to be building a stack yard, pulling up that little grade. Without warning, something happened that's never happened to me before, and I hope it never happens again. It set my heart racing.

We're mowing up that hill and Red does a crow hop in place, kind of jumping straight up in the air. I look up and the top hame strap, normally seated in the groove at the top of the collar on her harness, has broken. And the two hames have fallen down. The front of that hame assembly, with that strap, is cradled around one knee and she's trying to get her knee out of it. She's hopping around, they've stopped but she's in place, hopping around. I take the mower out of gear and lower the cutter bar for friction, because I know that mower is heavy and will want to roll back. And as it rolls back, its going to pull that harness in around her lower legs and we're going to have a problem; potentially a wreck. I climb off that mower, walk around as calm and sure as I can, talking to her as I come around in front.

Oh, what a mess. Oh, what a mess. I'm trying to figure out exactly what to do with it? The harness is down around her knees in the front. It's still hanging on her back from the back pad, it's still hooked to the mower, and the mower, even with the cutter bar on the ground, is dragging backwards, it's pulling up against her legs. Those horses are taking steps backwards slowly and Red is very nervous, and Mollie is starting to pick up on her nervousness. I'm thinking, "I'm going to have a wreck here!" And I'm trying to think what to do and then it hits me. "Mister, you are telegraphing how you feel, and your anticipation, to these horses. You're getting them excited by you being excited." So I try to put myself someplace else. I try to calm down and I succeed at it.

And I put my hand up on Red's nose and I talk to her. And if she tries to step back because the mower moved back a little bit, I let her step back. And I explain to her, "Red, this is kind of a tough spot. I'm not sure if I'm going to be able to do it, but I'm going to grab a hold of this neck yoke and I'm going to try to pull this neck yoke towards me. That mower's awful heavy and I'm going to need you to be working with me on this Red, do you understand." And her eyes are watching me, they are watching me intently. She doesn't understand what I am saying, but she likes the tone of my voice now, and it's already starting to calm her down a little bit.

I take a hold of that neck yoke and I pull up on it. Gosh it's heavy, because not only am I pulling up on the end of that tongue, but as I pull up on the neck yoke I'm applying just enough lift on the end of that tongue that the mower is wanting to roll back a little bit. I get my knee underneath that neck yoke and raise it up while I get those hames up on top of that collar. Then, with knee carrying load, I thread that piece of the hame strap in and cinch it on. I'm having to do that in-between these two horses and I know, if something should happen, I am in harms way. I'm right where I don't want to be, but I've got to get this all straightened out, there's nothing else I can do.

Left to right; Mark, Lynn, Jean Christophe, Prince and Pat. Photo by Kristi Gilman-Miller.

So I'm holding up that neck yoke with one knee and I'm reaching up around Red's neck and I'm whispering to her and I'm being as calm as I possibly can be. It's hard, it's heavy but I finally get it just snugged in there and I buckle it on down. I'm sweating and gosh my knee was hurting bad. I finally got that all situated and I get myself pulled back away. I got a cramp in my back and my knee was killing me, (it was black and blue for a month after that). And I look up and Red turns her head slightly and gently pushes her nose against my chest.

Quietly, I went back and climbed on the seat and swung the mower, got it on the flat and stopped, giving my universal command "whoa, stand". Back in front of the team again, I made sure I'd gotten all of my repairs right. I got everything functioning properly. And I realized that I had just come, after all these years, to intuitively understand two very important things. One was the geometry of that hitch, and what that mower was doing, and what I could and couldn't do to try and save that situation. And the other was that I telegraph how I am in those situations to the horses. I contribute to how they take that situation. Red gave me every measure of trust. It was a frightening situation for her, a very uncomfortable one, one that she wanted to get away from. But she had a choice and she chose to work with me. And when we made it all come out, she thanked me, and I thanked her.

Author mowing with Ed Joseph's horses. Photo by Kristi Gilman-Miller.

Chapter Thirteen

Comfort Diagnosed

Callie's Bad Day

"When I hear somebody talk about a horse or cow being stupid; I figure it's a sure sign that the animal has somehow outfoxed them"
- Tom Dorrance

Can you tell when a working animal is uncomfortable? Are you able to look at them standing and/or moving and say "they look comfortable" or "something's not right here?" Are you able to discern the difference between nervous, anxious behavior and clear discomfort? And what about those animals who are so dog-tired weary, not from just the day, but from ages of work, that they might appear to be comfortable when what they are is resigned? Are they telling you a lie? Even the most experienced teamsters might not be able to tell by a quick observation, or photo, but they darn sure can see it if they are on the lines with those animals for a period of time.

Jerry Hunter took this photo at the 2016 Horse Progress Days of a 4 abreast of Belgians. Look closely at the head set of each, their general posture in movement. These are four comfortable horses, well harnessed and matched to their work.

Many is the old teamster that will tell you that horses are honest and true. I agree. The horse won't lie to you. It won't make things up. Sometimes a really smart one, a clever one, will try to get away with something. But for the most part, for the important part, the horse will always tell you true. And the way they tell you is in the way they act, what they will do, what they won't do, what makes them nervous, what seems to make them comfortable. You need to be able to read those signs. You need to be able to understand those signs. You have to respect them, because if you don't, it will be a measure of how inefficient you are with working horses.

I have put up hay loose, in big haystacks, out on the field. I love the whole system, everything about it and there's one tool I really am enchanted by. I can't get enough of running the buckrake. It's very unusual, quite different from any other implement for horses. Imagine two horses, driven singly, they're 10 to 12 feet apart and they're pushing a big basket of wooden teeth in front of them. You're sitting all the way back, between two crazy-wheels at the back corners. And the way you make it go is you speak to both horses as individuals. You've got one line on each horse. That line will allow you to stop either horse independently, or both horses when you pull back on both lines. You stop one horse, and have the other horse go ahead, and you turn. And you can turn on a dime. I mean, that implement will spin around like a carnival ride. So, with time, the horses realize that you are asking only one of them to move ahead, or to back. Sharpens those communication skills with driving horses. I love going out and bunching hay and bringing it in to the stacker with a buckrake.

I was out doing that one summer day with Callie and Lana and we had a brand new harness. I was switching harness styles, I had some problems with the buckrake when using yankee brichen-style harness which put the brichen above the tail. It created some discomfort with the way the backing and braking system works on a buckrake. That's a little bit esoteric and I didn't really want to get too much into it except to explain that I got a brand new western basket brichen-style, harness. I put it on the horses, got it all fitted out and took them out. I hooked them one on one side, and one on the other side, to that buckrake and went into the field. It didn't take but five minutes, while we were pushing our first load of hay, Callie was just acting nuts, and she doesn't ever act nuts. She was the straightest, truest horse I've ever had. I mean I could snap a chalk line on the ground and she'll walk it for me. Now here she was wandering around, and turning her head right and left. I stopped, went up and checked the collar, and I checked the bridle and I checked everything. I was looking for flies and bugs, and I was looking to see how her breathing was, to see if there was something bothering her eyesight and I couldn't figure it out. She was just all over the place. So I went back on the buckrake and went to working again and I never could figure out exactly what was wrong, not until finally I decided, "this isn't going to work."

In these two photos, can you tell if the worried horses are uncomfortable or frightened or just brand new to the game of working?

Tom Odegaard drives his four abreast in North Dakota, photo by Fuller Sheldon. It's clear they are comfortable.

This mare that I had done everything in the world with: plowed with, mowed with, raked hay with, I had disced with her and harrowed with her. And now, I mean she had just gone stark raving nuts in the harness. She didn't want to pull, she didn't want to stand still, she didn't want to do anything, she just wanted out of there. I decided to use caution and unhooked her from the buckrake. Figured I'd take her back to the barn and go get Polly and Anna, Red and Molly, or some other team and hook them to the buckrake, so I could keep working. I took both Callie and Lana back to the barn, walked them into the stall, trying to decide whether or not I was going to yank the harness off of them or not, because I was anxious to get back to the hay. And I was along the left side of Callie, trying to make that decision, pulling her bridle off, getting her halter on her, so I could have her tied up to the manger.

I looked over, just below the trace, where it hooks into the hame, right there at the shoulder and I saw a little tiny rivulet of blood. So I lifted up the tug and saw, on that brand new harness, that they had used nails to hold the three-ply of leather together and one point of one nail was pushing through that tug and in that, just short, 45 minutes we'd been out in that

Ryan Foxley and his good big team turn the manure spreader on Littlefield Farm. No sign of discomfort. Phto by Joe Finnerty.

Lise Hubbe and her legendary roan mares return, comfortably, from spreading manure on the Scio, Oregon farm. Photo by Kristi Gilman-Miller.

field, that nail had ground a hole the size of a dime in her shoulder. That's why she was acting the way that she was. She was trying to tell me all along that something was bad wrong, and it was. And I was trying to figure it out, to my credit, but I never did - not out in the field. I guess I should have been looking under all those straps. I share that story just to demonstrate that these horses will communicate to us, and we should trust the message when they are acting out.

Jessica Lindsay plowing up a green manure crop with her handsome Mustang team. This young lady is representative of so many more young people round the world who, because of their successful beginnings, assure the future of the working horse.

The Charlie Orme clan, Belgians included, captured at an Ohio plowing match. Photos by Rick Conley of Lawrenceburg, IN

Susanne Burkhardt plows with Dean McCool's championship team at Yamhill contest. Photo by Kristi Gilman-Miller.

Pennsylvania Amish Farming scene captured by William Castle.

Handsome is as handsome does. Photo by William Castle.

Chapter Fourteen

Best Horses

Polly and Anna's Lesson on the Buckrake

> *"Horse sense is the thing a horse has which keeps it from betting on people."*
> *- W.C. Fields*

For four decades I've depended on horses to get my farm work done. And you don't get your work done if the horse gets its way whenever it wants. There are lots of times when you've got to exert yourself. You've got to put your foot down, not unlike raising children. My philosophy is that you stand in their way and make them choose the direction that you want them to choose, that way it's their choice, and you're not violating that rela-

A brand new buckrake and a young well-trained team with zero experience on the implement...

...allowing team to get accustomed to their working positions next to the buckrake...

tionship of trust that you are working towards. But there are times when you've got to use a heavier hand.

Polly and Anna, a pair of Belgian sisters, were only four years old when I first hooked them up to the buckrake. I've already raked and mown hay with them. My son Justin has worked them and my good buddy Jess Ross had driven them. Daughter Scout learned, with Polly and Anna, to drive when she was just four years old. And they were good dependable and very smart young horses. But the buckrake is an unusual apparatus and it takes a little bit of getting used to. Its not natural for the working horse to be pushing something ahead and in front of it, or to have what's in front of it follow when that horse backs up. There's quite a bit of confusion in the beginning. But I like to think that when a horse learns to work the buckrake that its a pretty clear indication of their intelligence and their acceptance and their trust in you.

Polly and Anna, as I said, were smart, willing workers, but at the same time not above trying to take advantage of the moment. In fact, even with as much work as she'd done, I'll have to admit that Anna had snuck in a few little kicks and two bites to get even with moments that she didn't really appreciate. She was a willful filly. And her team-mate sister, Polly, was pretty much sure that she was smarter than I am.

When we hooked them up to the buckrake for that very first time, these two mares stood about ten feet apart and independent of each other. In between them there was a framework of 2 x 2's and 2 x 4's running forward to that basket of teeth that would be pushing the hay ahead of us. A very unusual apparatus. I was sitting on the seat, dead center and behind them. We hooked them up, and I had a couple of people there to help me with any entanglements we might have. I needed that security because they were pretty much out of my reach, my being way back on the seat.

I spoke to them and they took a couple of steps forward and then stopped immediately. They didn't like that basket going in front of them. I let them stand there for a minute and then I asked them to back up. They leaned back and the minute that basket started coming towards them they backed up a little bit faster and then they just stopped, They froze up and just stood there. I could feel them get a little bit balky or stubborn. I understood it. For them

..proceeding with the hitching. Note that the hay buck basket is all in front of the horses which are to be driven with a single line to each animal. And when they backup, those teeth follow them back, an eery sensation for the first few times. These pictures were taken at Kenny and Renee Russell's Workhorse Workshop on their farm in Mississippi. This particular year Lynn Miller conducted a special event, building a buckrake and training a team to use it.

Kai Christiansen driving his Belgians on a buckrake at the Grant Kohrs Ranch in Deerlodge Montana. Photo by the author.

this was all wrong and they weren't about to get into a wreck. That's a remarkable, even mule-like, quality for any young team to exhibit. But I felt myself tightening up, this isn't right, "this isn't the way it's supposed to be," I thought. Took but a brief moment for me to assess the harness, hitch and equipment - all was right and good. No extenuating circumstances.

I was enjoying the moment because I knew that they were safe and intelligent horses. I knew they were going to learn this. Maybe two or three minutes had passed, I let them stand there that long to let them get used to where they were hitched and how everything was hooked up. Something changed in their attitude and they looked at each other across that ten foot expanse and, then, they turned their heads to the outside. Polly, she turned her head to the right. Annie, she turned her head to the left. And they tucked their butts in, towards the center of that buckrake and they both, in unison, as if they had rehearsed it, they proceeded to sit down on that buckrake! I didn't know what to do right in the moment but I thought, "wait a minute, everything's going to bust up here." So I lifted up both lines in the air and swung them down and hit those horses as hard as I could on their butts. I said, in a stern voice, "What the heck are you doing?!" And they stood up immediately and stepped out, and we went out and gathered hay. And ever since then, they have been one of the best teams on a buckrake I've ever had.

Kai takes a jag of loose hay to the waiting Beaverslide.

There are many lessons in that little story. Most important, if you are new to the craft understand that there are some situations with work horses that do not forgive ignorance or innocence. Polly, Anna and I got through that little learning experience because we knew each other, and we were, all three, experienced.

Another team, hitched to a forecart, pulls the beaverslide up. The loose hay dumps at the top of the slide, filling the rail basket to form a haystack.

From here..

...to here, all with horses.!

Willliam Castle of Shropshire, England, designed his own one horse variant on the buckrake. In this case the basket runs behind and to the side of the horse.

There is content in this book which, on the surface, would seem to fly in the face of political correctness. Striking your work mates to get them to pay attention, and to get them to straighten up, probably fits in there. In my defense I offer that I am old, I live a long way from media centers, and for me the concept of 'politcal correctness' is a four dimensional oxymoron - one which befowls itself in the process of negation. But that doesn't stop me from exercising my born right to appropriate hypocrisy. I say don't you dare raise your hand to your horses ... unless the situation clearly calls for it; AND you feel confidant that you know your horses well enough to predict the extent of their reaction. It is all about consequence.

Photo of buckraking at the Grant Kohrs Ranch in Deer Lodge, Montana, taken by the author.

"No philosophers so thoroughly comprehend us as dogs and horses."
~ Herman Melville

Chapter Fifteen

What's Possible

Haying at Night - Feeding in the Dark

One of the goals you ought to have working with horses is to get to the point that you can trust and appreciate them enough that you will let them help you get the work done. I don't mean just the business of pulling the load along, I mean using their

Ryan Foxley and team spreading manure in the half light. Photo by Joe D. Finnerty.

intelligence and their unique capacities and actually letting them help you get the work done. You get to that point and you're actually singing, maybe even singing in the dark.

Haying at Night

Many years ago I visited Arizona and the Cheatam Brothers Dairy. Found one man I could talk with and he took me on a tour of the equipment lot, explaining that the horse work didn't happen til after dark. It was 104 degrees so the haying work happened after dark and it was all horsedrawn. "Oh, I see. It's cooler at night for the horses."

"Ya, but that's not the reason. It's too hot and dry. Alfalfa wont hold together to bale in the day time, turns to dust. We need some moisture to hold it all together."

"How do you see to do the work?"

"Oh, we have some lights. But the horses don't need them. And we work when there is a moon. We drive four abreast on these wing-rakes and four abreast on the balers."

The engined New Holland balers were rigged with heavy single crazy wheels under the tongue. A seat was mounted there and, in front of it, a dash board of sorts with clips along the top rail.

"Those are for the lines. We clip the lines there 'cause we walk back here alongside the engine and the knotters."

"We hay four hundred tons. Running twenty head or more of draft horses. Pretty much making hay all summer long."

Dairy likes it this way 'cause labor is cheap. Short story.

Feeding in the Dark

Ed Joseph used to work for the state highway department while farming with his horses. At three in the morning, he and one of his adolescent daughters would go out in the dark and harness one of their Belgians and hitch to a stone boat. They'd hang a battery light on the hame, load hay bales on the stone boat and make the rounds feeding cattle, horses and the donkey. They were done before sunrise. Did this every winter morning, often either rain or snow. Imagine the memories those girls have stored up. Ed let them drive while he pretended that throwing the hay out was hard work. The end result was that those horses, Pancho, Meg, Peg and Kit, had Ed, Charlene and Nathalie pretty well trained.

As these lovely Australian photos will attest, sometimes working horses can feel almost primordial, as in coming from, or reaching back into, dark recesses of subsistent human adventures.

Littlefield Farm Suffolk mare at early morning barn cleaning chores. Photo by Joe D. Finnerty.

Chapter Sixteen

Today and Tomorrow with Work Horses

Innovations / Applications

Can't get enough of it. Time with the horses I mean. No matter how long I've been at it, I just can't seem to get enough. And I'm not talking about watching them in the pasture or standing around with them (something we old people do far too much of). I'm talking about working them, working with them. After four plus decades it's still what I want to do. And, today in the twenty-first century it's easier. Today we have people to share our adventures with and companies making the implements and gear we need.

Today I can't escape the feeling that we're supposed to at least try to contribute constructively to growing and protecting the work horse and mule way of working. In and of myself, I

William Castle of England with four Geoff Morton Clydes.

Jess Ross on Singing Horse Ranch, running his favorite implement, the buckrake, with his Percheron team.

don't have much to contribute, ideas and answers I mean. I once thought that I did but I've matured, grown up if you prefer, and now I see I'm supposed to collect the information and pass it on (ergo this book). What I might think about any of it is just a silly drop in the vast bucket. Clarity and useful knowledge don't always ride in the same box car. That's where I'm at now, with horses and with my philosophical leanings. It's all clear to me, just don't ask me to explain it. I don't know much, but I can play the instrument. And I wish more people would join the orchestra.

Two of my many teamster friends, Mike Atkins and Jim Butcher both of the Ohio regions, agreed with one another a few years back that their only regret with the choice to work horses and mules was that they hadn't started sooner. An unidentified man sent us a note saying that life wasn't worth living before he started working his good horses. Walt Bernard of Oregon told me, over the phone, that when his good Percheron mare passed on he felt his whole world come apart. Right up until his death, my buddy Jess Ross's best moments were peppered with memories of teams he had worked. The cavalcade of examples goes on forever. Each and every great teamster I have known, and there are hundreds I call friend, each and every one, though they employ different techniques and prefer different animals, share one important thing in common, they love this work. But all this is too personal, we need to sideways the conversation to something a little more objective and clearly spoken.

If our goal is to advocate and preserve and improve animal-powered agriculture, where do

Anne Nordell cultivating near Trout Run, PA.

we start? Let's begin by identifying the social environment we must work in; let's look at the public perception.

Why farm with horses and/or mules?

One public notion: Within such an archaic concept as animal-powered agriculture, can there be any practicality for the twenty-first century? After all, this is something they did a hundred or more years ago. Why on earth would anyone in his right mind even try it today? Can you get across the field? Can you get your crop in? Can you find the implements and gear that you need? They aren't making this stuff any more, are they? Would anyone in their right mind choose such a way of working?

It's a subject worth talking about, today. But perhaps that is after the fact because, though it may not be apparent to the uninitiated, hundreds of thousands of folks ARE already depending on horses and mules in harness to get their work done - TODAY! The photos selected for this book point to that. (I had to choose from a mountain of pictures and even with a book of this size, space is limited.) Surely there are enough images here for you to get the idea that the teamster's craft is alive and thriving.

The middle of any discussion about the future of animal-power must be some sort of un-

Gerry Lee of Estacada, Oregon, with his three Hackneys hitched to a walking plow. Gerry has been a fixture in draft horses, of all sizes, for all of his life. Logging, farming and pulling have been his passion. He, and his extended family, help keep the traditions alive. Photo by Kristi Gilman-Miller

derstanding of who's doing it. And with that, a look at what drives them to do it (no pun intended). The most interesting and important distinction here is that most of the people who are working in this fashion have made the conscious choice to do so. They want to do this and they enjoy it. They don't have to work this way. This is good news for most of the animals. They just might be better cared for. It also suggests that these people have a real stake in making the system work better each day. And these folk come from every walk of life; every economic strata, every level of education, every religious background, every ethnic construct. Some come with tight budgets or no money at all, while others can afford to spend just about anything it takes. Some come with agricultural educations and engineering backgrounds while others don't have a clue about soil ph or hydraulic couplers. Some come believing that God wants them to work this way while others come with the delightful confusions of a secular romanticism. While there are many new people involved with draft animals, a great many have been at it for a long time. Each and every one of these good people come to this notion of animal power because he or she are attracted to it.

Prototype single row cultivator by Vitimeca. Found in Europe. Photo by William Castle.

Detail of the adjustment of the European Equi-Idea plough. The screw at the back pivots the body in relation to the beam, so with the wheel at a fixed height from the beam, the angle of the plough body to the beam, and therefore the depth of the plough, can be adjusted. The screw at the top allows the handles to be quickly adjusted both up and down and from side to side. Photo by William Castle.

Jelmer Albada in 2013 working a Belgian horse on the Mellotte tool carrier at Roxbury Farm in Kinderhook, NY.

Eric and Anne Nordell operate their market garden in Pennsylvania with horsepower.

Why farm with horses and/or mules? Because the animals work and because we can do it. Why farm with horses and/or mules? Because they fit so sweetly into the well-rounded, diversified, mixed crop and livestock family farm. Because they are self regenerating. Because they keep the energy equations where they belong, on the farm. Because they can be wildly profitable. Because at the end of the day they make the tired, satisfied farmer glad to be alive. Because, for the right people[1], they are the most exciting, efficient, and rewarding renewable motive power source for agriculture.

---

1. Not every person is cut out to work horses or mules. And not every equine is suited to work in harness for people. This way of working is a highly subtle, functional craft requiring initiation by immersion for the prospective teamster. There are many workshops, clinics, and demonstrations around the continent offering outstanding assistance to any who are considering working horses, but they are just that, an assistance, and they will never replace the time required actually doing the work. Books, magazine articles, and videos are important aids but the information they contain is constrained by the media format and frequently offer a severely cropped viewpoint. There are many formulas floating around which suggest that all you need to do is follow this checklist and your time with the working equine will be safe and easily understood. That is most definitely not the case. The beginning horsefarmer is best served by at least one apprenticeship or internship with an experienced teamster, where extended time with driving lines in hand is available. After that, success with the system may depend on having knowledgeable practitioners available to lend a hand at key moments.

The Nordell's have authored a long-running and vitally important column in Small Farmer's Journal entitled "Cultivating the Bio-Extensive Market Garden.' This column documents their organic farming practises as well as of those employing similar techniques. These photos depict the diversity and cleanliness of their farming. Eric is one to say that he farms so he has an acceptable excuse to work his horses. They have sustained their successful operation for more than three decades. As one observer commented "the only criticism of the Nordell's is that they make it look a whole lot easier than it is."

Thomas Roberts of Lenoir, NC, cultivating corn. Photos by Paul Roberts

Jessica Lindsay, above and below, deep in the heart of horsefarming in Dexter, Oregon, with tractor GD rake and Marvin Brisk-built Culti-mulcher. The horses belong to Walt Bernard.

But they are just a motive power source, a way to pull implements across the field, the wagon down the road, the log through the woods. Their relative efficiency can be dramatically affected by the vehicle, by the implement, by the harness and hitch gear. (And, of course, by the imagination and polished skill of the individual farmer. By the artistry.) We in turn, as horsefarmers, may also be greatly affected by the tools which are available to us, whether they are commercially manufactured, shop-built, or those lovingly maintained older implements. We, by necessity, are always looking for the better tool, the finer tuning, the lessening of draft, the greater comfort for our animals. Experience has taught us how such steps, small or large, extend the practicality of this way of working.

Implement Design

Forty years ago there were NO commercially manufactured horsedrawn implements widely available. None. Zero. Zip. Anyone who might doubt that there has been a serious resurgence of interest in animal power only need be shown the evidence that today many new and various shops and small factories are busily manufacturing an ever growing lineup of

Paul Schmit of Luxembourg modified a cable-lift, Trebbiner trailer to pull behind his forecart. These container trailers were originally designed to be pulled by cars or light trucks.

Both in Europe and in the U.S. new horsedrawn farming implements are springing up every year, with balance, efficiency and best draft all factored in to advantage. Above William Castle captured this picture in Germany of a ground-drive forecart working a PTO mower. While below, Doug Sheetz photographed this Shipse Culti-mulcher at a Horse Progress Days event.

After long days in the field with your work animals, the relationship is so strong and positive that many folks take it beyond to use them for relaxation, fun and pleasure. The "Chukker-Mobile" Bob Whitter driving "Chukker" from 2nd sled on the left. Photos from Dick Brown.

outstanding, sophisticated implements and attachments. (There is no better place to see this than the annual Horse Progress Days - set on a revolving venue in Amish communities in Ohio, Pennsylvania, Michigan and Indiana - with thousands in attendance.) And all that (the manufacturers, the new implements and the event), has sprung up over the last forty years. Every one of those recent years has seen a steady growth in sales for these new implements. We are in the midst of a true rennaisance for practical horsefarming.

Yet, as is so frequently true of periods of rebirth and rapid growth, there is a wide open, almost gold-rush aspect to this resurgence. It's as if everyone is poised to jump into the fray, go to their farm shop and break out the welder to start shaping the next best thing in animal-powered tools. They may feel that there is opportunity for almost overnight success with the invention and construction of specialized implements and applications. And within this

small world of inventor's euphoria, there is some confusion. Not everyone understands what is needed, let alone possible.

Universality and adaptability are essentials for this blossoming spectrum of farm implement innovation. "This button-on plow would be a lot handier if it could attach to that forecart or that one over there."

The Yard Hitch cart hooked to an I & J plow at Horse Progress Days.

The Wengerd family, of Pioneer Equipment in Dalton Ohio, spent years developing this outstanding small farm implement based on the old original straddle row cultivators but employing modern materials, leverage designs and an array of quick attach implements that go from plowing, to tillage, to planting, cultivating and even some harvest applications. It is called 'The Homesteader.'

Below: Recently William Castle photographed this new forecart design and rake at Germany's Pferdstarke event.

New machine by the French company Vitimeca, made for loosening the soil in vineyards. Photo by Williiam Castle.

and "Why is it that this motorized forecart doesn't offer a way to hook up to and operate that cane berry sprayer?" or "Why is it that this forecart system doesn't lend itself to working between or over these rows?" Decades of manufacturing in the tractor sector has, of course, paradoxically tended to reduce adaptability of implements to a workable set of variants. And those, be they hydraulic coupling systems, catagory-one three point hitches or power-take-off hookups, seem to be somewhat perfected within certain scales. Obviously huge tractors don't marry up well with most row crop tools and small cultivating tractors aren't able to attach, let alone pull, major harvest and tillage implements. But within those scale parameters there is a definite paring down to common hookups. We don't yet enjoy this universality of "plug'n'work" inside of animal drawn systems, with Pioneer's Homesteader being the notable exception.

It may come down to "first things first." Within the world of animal-powered implements

14 year old Lewin Ahrens, with his family's Fjords and a home made 4- wheel, ground drive forecart and a 1960's self loading forage wagon. Unusual breast collar harness featuring neck straps to carry tongue. Photo by William Castle taken at Pferdestarke.

Mac Mead at Pfeiffer Center discing a raised bed with a team of Haflingers. Photo by Nathaniel Mead.

This interesting machine is for working between vines. The bar in front of the mini plough body moves against a spring to move the plough inwards as it passes the vine. The sweep at the back similarly moves inwards when it meets an obstacle. Caught at Pferdestarke by William Castle.

and procedures, horsefarmers must have a high regard for those engineering aspects which respect the essential need for ever greater balance, lessened vibrations, smoother operation, and improved draft. This has to be the higher priority. We are asking horses and mules to pull these implements. The quieter, smoother, and easier an implement works the more work we will get done over the long haul. The second priority has been for how it all might hook together. Here we find ourselves still dragging fanny. But we know it and work is being done to connect the points in the best and most universal fashion. We find ourselves, as an industry, in a little race to perfect horsefarming implement systems. And the winners of that race will be the manufacturers who pay closest attention, and due homage, to the in-common mechanical needs of the widest array of farming procedures. Simply put; we need it all to hook together and we need it to work well for us.

Speaking of needs; for the manufacturer it will always come down to that primary question which must be asked of as many horsefarmers as possible - 'what do you need?' The best design imaginable will have a short life if no horsefarmers pick up on it. The not so distant past has seen many examples of just such phenomenon. The equipment inventor who works in a

German Elmar Stertenbrink drives his Dutch Drafts hitched to a limer, a special construction for forestry work. Photo courtesy Erhard Schroll, Starke Pferde.

I & J Cultivator at PA Horse Progress Days, photo by William Castle.

Foliar feeding all vegetables weekly with a five-row boom sprayer. David Fisher and Anna Maclay of Natural Roots Farm in MA.

vacume, listening only to his or her own excitations about design and function, will usually fall way short of exciting the farmer/customer. If it doesn't sell, it doesn't survive.

THE FARM LABOR FACTOR

New/old pressures present themselves with increasing insistance in farming circles. Society across North America, numbed as it is by shallow education, spiritless electronics, synthetic foods, and mood-altering drugs, is less and less inclined to contribute people to work on farms. Whether in the Maritime Provinces of Canada, or the deepest canyons of the Southwest, from the remote woods of British Columbia to the swamps of the gulf-coast South, or dead center in the Midwest, the same sad song is heard - "we can't find anyone who wants to work on our farm." This is true for all brands and persuasions of farmers, from the Amish to the nonAmish, from tractor to horsepowered. This applies specific pressure when it comes to the equipment question. When a horsefarmer is asked, today, what do you need in the way of farm implements, it is likely that his or her answer will be colored by the resignation that they will have to do more with less labor. The outsider might wonder at such a predicament. "Well, if labor is the problem, why don't they just switch over to tractors?" Most dyed-in-the-wool horsefarmers would never think of such a thing, they work horses because they

Paul Schmit (of Tuntange, Luxembourg) and Schaff mat Paerd field testing the new SmP Sei-Roll 1.0 for horse traction, the Seeder Roller. Schaff mat Paerd's studies regularly appear on the pages of Small Farmers' Journal, and may be seen either at www.smallfarmersjournal.com or www.schaffmatpaerd.org.

In Northern Italy the two agricultural machinery manufacturers MAINARDI A. s.r.l. and REPOSSI Macchine Agricole s.r.l. produce a vast range of haying equipment with pto and respectively hydraulic drive, also ground drive hay rakes.

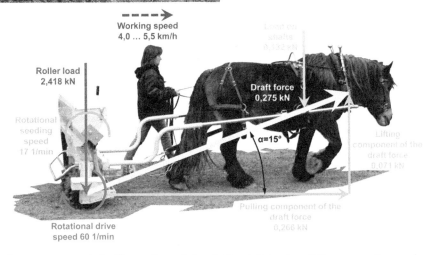

The non-profit association Schaff mat Päerd (with Paul Schmit of Tuntange, Luxembourg and Albana Moscardo of Verona, Italy) support animal-powered agriculture and the development of new equipment with exhaustive testing and related studies and publications. The photos on these two pages are from their work.

Geoff Morton's 23' wide flexicoil gang pulled by four horses in Great Britain. Photo from William Castle.

love it, because it is a way of working that is in their blood. Many of them have just enough outside experience to know that tractors don't always represent less labor. Work at something you love - is that labor? Work at something you dislike - every moment is labor. Well, the work still needs doing, we have a crop to get in. If it is just mom and pop to do the job we need to think about how certain implements might help us out. It's a paradox. We love our way of working, we love our labors, yet there's a need to find labor saving implements just so we can get it all done.

Or do we? I was reminded a few years back, while helping out at the McIntosh Threshing Bee - helping operate the binder and pitching bundles into the thresher - that small farm communities used to be knit together by working social events where neighbors met to help neighbors. That weekend we all had a ball and ate way too much. These "bees" or work parties made and make some difficult tasks more than bearable. In fact, it was common in the Pacific Northwest for several neighbors to go together on the ownership of big implements. My buddy, Ed, the week before the Threshing had driven 3 hours to lend a hand with my haying for two days and we lamented the fact that we don't live closer together so it could go both ways more often. Ed went home for a couple of days and then returned to help at McIntoshes. Maybe one logical solution to the labor problem is to pick and choose farms near folks of like mind and work. But we must accept that for some people this is not an option.

Walt Bernard's GHaw tool carrier cultivating a farm in Dorena, Oregon

For them they need their horse tools to save the most amount of hand labor.

It is a balancing act, one which may offer the insightful, imaginative, equipment designer complex new possibilites. Here we have a craft, horsefarming, which is dependent upon human involvement and which is loved by its adherents because it is a craft and not an industrial process. Because of the shortage of labor, some of these craft-based farmers are forced today to look into higher tech implements, employing industrial production aspects, to get the work done. How do we balance craft with industrial process? Does the classical violin maker forgive the assembly line in any part of his craft? Does the passionate horsefarmer forgive the internal combustion engine in any part of his craft? Come up with tools that get the job done while honoring the craft-based nature of this way of working and we horsefarmers will beat a path to your door.

Finding where to look

Back around 1971, when I first took to the idea of working horses in harness, people rightly wondered how hard it would be to get equipment. No new implements were being made and shipped about. Most all harness was being made by either Amish shops or for the show trade and it was leather. I recall that in my beginning with this craft an early difficulty was finding where to look for the good serviceable older implements, and where to find information on fixing them up. (Cracking that mystery actually led to the birth of the Small Farmer's Jour-

Suffolk mare running a barn scraper at Littlefield Farm in Washington. Photo by Joe D. Finnerty

nal, but that is another story.) For today's beginner, knowing where to look is still a challenge. Moreso today than back 45 years ago, the "where" has become a moving target. In the nineteen seventies you had to learn to sniff out nearby forgotten regions, old sheds and weed rows to find the best implements. I remember a fallen down ten by twelve shed buried

Ed Hamer runs a Shire team on a cultivator in England.

under blackberry vines near the then rural Aloha, Oregon where Ray Drongesen and I had been led to believe there might be a serviceable old mower. We had to cut back vines and pull free boards to look inside, and sure enough there sat an old Champion mower with original paint and looking like it might have seen only one season cutting. It cost Ray $40. Another day Ray got a phone call from a widow woman who said she had some sort of old farm tool in her barn and wanted it hauled off right away. We went right over and found a pristine McCormick Deering Hay loader. She wanted no money, just take it away. That's how I came to own my much loved and used hay loader. Over time these local finds diminished, the tools were scrapped or antiqued or found by other horsefarmers.

Then came all the regional horsedrawn equipment auctions such as the one we used to do every April (old age retired it) and a new breed of traders were invented. Men and women who travelled around the country looking for deals on old implements and putting together loads to take to specialized sales where, on average, more money could be made. This meant the tools sprouted legs. Now it was possible for someone to purchase a mower at an Amish sale in Ohio and consign it to the big Waverly Sale in Iowa where it would be purchased by a trader who took it on to a farm museum in New England where it sat for fifteen years until it was sold to another trader who hauled it out to Colorado to the Troyer Sale to be bought by a North Dakota farmer who, on a visit to family in Arizona sold it in the Four Corners region only to have it end up on a loaded trailer and heading for the auction in Sisters, Oregon to be purchased by someone who shipped it to their farm on the island of Oahu. This is what I mean by 'the where becoming a moving target'. None of this would have ever happened

Jessica Lindsay using the ground-drive baler to bale oat hay with Walt Bernard's draft horses in Dexter, Oregon.

without the resurgent interest in animal power. Now these old original tools have value. They haved value to traders because they have value to us who need and use these tools. It's a firmly established circle which ironically has depended upon relatively cheap fuel allowing all that travel and shipping. (We'll have to wait and see what 4 or 5 dollar diesel and gasoline will do to the circle.)

The bridge to newer implements

Everything I acquired back in 1972 was old: mower, rake, plow, cultivators, planters, manure spreader - and affordable. I built my own forecarts from old front car axles by welding shut the auto steering and bolting on a platform for a seat, a tongue, and a hitch bracket. They worked, but they were frequently out of balance, too heavy, too big or too small. It was a long learning curve before I came up with my version of the three wheeled forecart, featured in the Work Horse Handbook. That cart allowed me to more readily hitch up to pull-type ground drive tractor implements. This opened up a much wider range of possibilities for working tools. But the bug had been firmly planted, I had fallen in love with those older, direct application, horsedrawn tools.

Those first years of working with the original implements, the beloved number 9 mower, the Oliver sulky and gang plows, the McCormick riding cultivator, the wooden box Internation-

Horse Progress days, rakes demonstrated. Photo by William Castle.

al manure spreader, the John Deere corn planter, those first years felt akin to learning how to play an array of musical instruments. A great deal of finesse and balance went into designing those pieces, and into operating them as well. Sitting in the basket of the straddle row cultivator, feet in the steering stirrups, sighting down the tongue between two good horses, I had a sense I was farming, truly farming. You were down where you could see, smell and even hear the earth. Gentle efforts with your feet and you could direct the cultivation to avoid taking out plants. Each tool had its quirks and its perfect applications. And I, for one, learned early on that if they were to be kept in the field, I had to do two things. One, I had to learn how to maintain and service them, because there was no local blacksmith or antique implement repair shop. And two, I had to keep a close eye out for the same makes and models of implements to purchase for spare parts. Lest you think I complain about this aspect of keeping the horses working, let me say that I have always loved this part. I love working with the

Pennsylvania Amish farm scene, photo by William Castle.

Six Pennsylvania mules pulling a ground-drive baler at HP Days, photo by William Castle.

tools, knowing how to fix them, and I love the hunt for spare part rigs. And I know I'm not the only one. There are thousands of us old-iron buffs out there still crawling around in the fence rows and at local auctions.

There is so much that is remarkable and beneficial about the history of Horse Progress Days. But there are also problems, especially when the focus slips. Doug Sheetz took this picture at a HP Days where dealers influenced the organizers into allowing that a massive rototiller be pulled by an equally massive deisel-powered forecart and four belgians. The PTO powered rotary tines of the tiller provide ample forward thrust leaving the horses to walk in front and steer the apparatus. A sustainable power source married to carbon spewing heavy-metal disease?

A complex cart and seeder innovation displayed at Pferdestarke in Germany, Picture by William Castle.

How does this fit in with an industry, horsefarming, which continues to grow? Some people wonder about the continued availability of the older implements. Just as they wonder about the practicality of the older procedures. They measure picking up loose hay with a hayloader and wagon and stuffing it in the top floor of a barn against using horses to make round bales. They measure plowing an acre or two a day with a walking plow against four plus acres a day with a new gang plow. They see jobs they might do today with new implements that have no counterpart in the older tools, such as plastic-mulch layers, transplanters, and raised bed formers. And therein lies an interesting paradox; how to marry craft to some appropriate level of industrial procedure without killing off the craft aspect? An interesting paradox and a critically important question because if we don't answer it we'll stand by and watch as horsefarmers will opt for tractor setups to get certain jobs done. And that is a problem because those of us with experience know that when we have to use the horses to get a specific challenging job done the horses benefit in their training and conditioning. But when we leave the horses in the pasture and go get a tractor and baler, or such implement, we lose a part of why we are at this game and the horses lose valuable time in harness and in the field.

Sometimes this paradox is answered by the very working. Take for example the specialized needs of market gardening. Here is where the training, courage and intelligence of the animals and the teamster shine. A slow, patient walk is required of the animals as is the temperment that will put up with sheets of plastic mulch, seedlings, various sprays, and close quarters. Finesse is the name of the game. Here is a place where craftsmanship may still reign supreme.

Three bay mules on a new style corn cultivator. Pennsyvania Horse Progress Days.

Sometimes, this paradox is answered by those closest to the process. Today we have a couple dozen or so outstanding manufacturers of all new horsedrawn equipment, each and every one of which is connected by tradition, experience and heritage to the actual working of horses and mules. All of them understand what it means to depend on the harnessed equine. They saw needs and worked to fill them. Because they were successful in the beginning new variants, improvements, and implements were continually being added until we have what is today an outstanding lineup of the new tools. Today we have manure spreaders, liquid applicators, sprayers, broadcast spreaders, forecarts of every stripe, plows, harrows, cultimulchers, discs, raised bed makers, plastic mulch layers, transplanters, and so much more - all new. All because we, as horsefarmers, need them.

Some of the most successful new implement designers and manufacturers are, I believe, where they are because they understand not just what it means to hook a horse to the tool and get the work done, they also seem to intuitively understand intrinsic aspects of elegant design. White Horse Machine of PA, would on the surface, anyway, seem to have zero concern for the unique beauty of their implement designs. They obviously strive to make the implement work. But, taking their ground-drive Hydraulic Forecart as an example, they succeed on every front. Here is a tool which uses the motion of the drive wheel to pressurize a hydraulic cylinder. And that hydraulic force is made available, to a variety of hose outlets and

*Handsome Belgian mule team on walking plow
at Mt Hope, Ohio HP Days, photo by author.*

a cart steering function, by elegant manifold and operative handles. To see the cart is to see function. It works, everything is where it needs to be, it seems simple to a fault. But simple it is not. It was by necessity designed, and it is by application and directness beautiful.

It may be purely a philosphical discussion to invite argument by saying that the best design follows function and results in elegance. I would point to Pioneer Equipment's sulky plow equipped with the Kverneland bottom. Here, once again, the focus is directed towards function but the picture is one of elegance. It goes without saying that if Pioneer's first priority were to make a pretty plow, functionality might have been lacking. There are doubtless examples, but for this author most every implement that works well has its own inherent beauty. And beauty in design and function gets us closer to that matchup of procedure with craft. And it is that matchup that we need be most interested in. For you see there is always the risk that the evolution to more sophisticated tools will actually strip this way of working of much of its attractiveness.

I believe that one of the things we enjoy about working horses is that it allows us to be close to our work, to sit on tools that have us directly in and over the procedures. If more of our implements separate us from that working, in the way that a complex motorized pto forecart might, there is the risk that is will be less attractive and less accessible. By accessible I mean that the newcomers to this system may feel that the riding plow and traditional horse-drawn mowers look and feel like something they themselves could and would do, whereas a 45 hp

A rotary cutter for green manure crops. Photo by Doug Sheetz.

diesel forecart hooked to the complicated new round baler will seem daunting in the least. And that is saying nothing of the relative cost factor. Keeping this in mind, and holding out for designs and engineering that truly respect the best nature of the craft, new implements can be invented that will keep the attraction and accessibility alive. Doesn't that fit inside of the notion of elegant and useful design?

Please don't mistake this as a wholesale argument against motorized forecarts and round balers. I am not the one to make any determination on that. I am, as I said in the beginning, trying to gather information to share for us all to make the best choices and conclusions together and separately.

horse progress days interviews

Years ago, at the Horse Progress Days event in Clare, Michigan, I was involved in a filmed set of randomly selected interviews with folks, asking each of them what they were looking for in horse-drawn implements and how they saw the future shaping up. Those interviews were a real eye opener. The people we spoke with covered a very wide range of interests and experience levels; from an east Indian scientist to an Australian doctor, from a Thunder Bay, Ontario horsefarming couple to a young man thinking about a farm in Maryland, from Minnesota to Nova Scotia to California and beyond, from a succesful one horse farmer to a man who gets his greatest pleasure with big field hitches of nine and twelve head. From dreamers to doers and back again. And the main tone of the answers was practicality. The effectiveness

Ohio plowing scene at Horse Progress days. Paul Coblentz and his paints. Photo by author.

of the tool for the job needing doing and the cost[2] whether purchased or homebuilt, these topped the list. Almost half of the respondents attended HPD for information and ideas that might assist them in tweaking the tools at home or in building a new one. They came with

---

[2]. *Among the respondents, "cost" was a wildly relative subject. To some the cost of any new or used implement depended solely on the value married to the need. If they had work this tool could do, and if they could justify the expense of the tool to get the job done, then the value was there. The issue of affording the tool, for them, came down to justification. If they could justify it they would find a way to afford it. To others with tight budgets cost came down to what they felt they could afford regardless of value attached to need. If they couldn't afford it, they wouldn't consider it. There were a handful who were so enamored of the implements that budget and justification simply flew out the window. They just had to have them.*

Cultivating and fertilizing corn all in one pass with three Percheron horses. Ohio scene captured by the author.

measuring tapes and notepads. A few of them were so impressed with what they saw demonstrated that they made the leap to purchase a new tool right then and there, figuring that the value was apparent and that they couldn't improve on the design at home. With one qualified exception; a gentleman who believed that nothing new could compare with the old original tools, everyone else was favorably impressed with what they saw demonstrated in the field trials.

I have attended several HP Days, in several locations. And I have heard one common concern; that the show was the same each year, that nothing much was changing. If you saw one HP Days, they say, you saw them all. Perhaps from a distance this could seem to be the case, but up close I can tell you that each year has had notable and important variations. One of the most significant contributing factors to this stellar annual event has been the crowd of spectators. Thousands of people attend and from all over the world, with each year bringing in a different batch. The spectators with their interests, questions, excitements, and their purchases have a tremendous effect on the event. They literally shape what might occur the next year.

Pequea manure spreader on forecart and Shipse Cultimulcher, both demonstrated at Arcola, Illinois Horse Progress Days. Photos by Doug Sheetz.

Back to the interviews: In keeping with concerns for effectivenss and affordability, some of the folks were very keen on doing their own calibrations as to how many horses. They have seen the pictures, for example, of four abreast Belgians and Percherons pulling motorized carts with balers and haybines, the White Horse two-way plow, and soil slitter, and the Shipse Farm Supply cultimulcher. What they wanted to know, and preferably see for themselves, was whether or not two or three horses could do any of those jobs, or were four head necessary. They read where a 47 HP forecart was used on a round baler but was that abso-

Four Amish Percherons on a tractor wing rake.

Four Belgians pull a White Horse hydraulic subsoiling plow employing the WH hydraulic accumulator.

lutely required? Could they use a 24HP cart with any particular model of round baler? What we then found, full circle, when we passed on these questions to HP Days organizers was that they were anxious to answer them by demonstration.

On the 'how many horses?' front; there were for twenty to twenty five percent of the folks persistent needs for implements, attachments, modifications, and gear for single horse farming. I & J Implements may be heading that way as well as Pioneer Equipment's whole new cultivator frame as toolbar approach for the small market garden operation (The Homesteader see page 254). There are excellent and serviceable manure spreaders, sprayers, cultivators, plows, and such already available but something seems to be lacking. When you point these tools out to the access hungry single horse farmer, the response seems to be 'yes...but...'.

Another point which came out in these interviews was this issue of access but in a larger and more decentralized way. People from all points of the compass around North America travel hundreds and even thousands of miles to Horse Progress Days to see all these new exotic

Betsy Whittick of coastal Washington discs the grape rows.

implements BECAUSE they can't see them back home. Not unless they have a neighbor with one or two. HP Days is clear proof of the need and the demand for the tools. But our industry is still so young that distribution remains ridiculousy stilted. Most of these more sophisticated HD implement manfacturers are located in a handful of states; PA, OH, IN, IL, IA, MI, MN, NY. If you are a horsefarmer in Saskachewan or Georgia, California or Nova Scotia you have to travel a long ways to the factory, or to HP Days, to see and purchase the tools. This is holding our little industry back. It will change with time but it might change quicker with a little bit of coordinated internal effort. Most of these cottage industries have worked out the bugs for small-scale assembly-line production but they haven't even considered what it means to wholesale units to a wide spectrum of potential dealers.

There are quite a few of us horsefarmers but not nearly enough in most localities to warrant a specialized dealership of implements. Using important examples such as Meader's Supply, which serves New York and New England by stocking anything for the draft animal enthusiast, we might expect to see in the future such outlets in every region. But not for a good while. It makes me wonder if we don't have general farm supply stores in place already, offer-

Dick Brown combining oats with a JD 30 combine powered by Pioneer motorized forecast. Modoc Herbert the stallion is one in mane. Photos these pages by Robert Mischka.

Dick Brown spraying with Pioneer 20 hp motorized forecart.

Dick Brown picking corn with 1910 McC corn picker. Fred and John Dinnes with Belgians Below, Dick Brown with four head on motorized baler and wagon.

Corn harvest scene, Dick Brown and company. Photo by Robert Mischka.

ing tractor, livestock and garden supplies which could be encouraged to add 'horsedrawn' to their lineup. Something worth considering. But to make it happen there would need to be changes to how the individual implement manufacturers treated wholesale versus retail. The makers would have to work with the dealers to mutual advantage otherwise there would be very little incentive to carry the items. It is easy to see that a wider distribution of the tools would have an extremely positive effect on the increased growth of the work animal industry.

Wrap up

All indications from the equipment front are that we can expect continued strong growth within the horsefarming community. But the manufacturers must keep listening to the farmers.

We'll never see 75 cent farm fuel again. And it will take a century or better before we see a commensurate 400% increase in wheat and corn prices. That means hundreds of thousands of farmers will be looking for motive power options. Couple that farm field reality with the worldwide need for the best food, and I think you just might agree with me. If each and every one of us never misses an opportunity to improve the "work, look and feel" of horsefarming, the next thirty years should yield amazing results.

Two haying scenes from Italy. A ground-drive loose hayloader conveys the hay forward into the wagon.

Four Percherons fan around hitched to one of the new versions of a cultimulcher demonstrated at the 2016 HP Days in Indiana. Photo by Paul Hunter. These horses are all in sidebacker harnesses.

A modern farming scene from Italy showing a single mid-sized draft horse hitched to a dump rake and raking hay between the orchard rows.

Oregon's Lise Hubbe and her big hitch mentor and friend, Dick Brown of Iowa.

Lise Hubbe pulls out of the furrow with her seven-up of Belgians and Brabants. Her farm is in Scio, Oregon

This donkey hitch pulling the broadcast spreader was captured at the 2016 Indiana Horse Progress Days by Jerry Hunter.

Field work on Natural Roots Farm in MA

These photos are shared by Noel Vinet of Sainte-Clotilde, PQ and portray sugaring scenes in the north country "bush." Horses excel when it comes to the wintry woods work of collecting sap from Maple trees. Whether in Quebec and all of eastern Canada, or the New England states, the tradition is old and rich. A few years back, I visited Spence Farm in southern Illinois and was surprised to find an equally strong maple sugar tradition in that climate zone.

When the weather is cold and there is snow on the ground, a good team of horses can be a delight to move feed, firewood, logs or maple syrup. They'll start in the cold, and, with proper shoeing, handle most traction issues.

One winter, while I was feeding with a bob sled similar to the one above, it was 40 below zero. I took frequent breaks to stand between the horses up front to warm up a little, I think it may have saved me.

In the winter time, when feeding everyday, it is a good idea to keep the bridles in a warm place. It willl pay dividends when it comes time to put that bit in the horse's mouths.

Bobby Jones, above, and the entire gathering, below, plowing at Mike Down's farm for the Kentucky Spring Field Days a couple years back.

Mike Atkins had broke several of these horses for Charles Orme. Photos this page by Rick Conley.

Tom Odegaard of North Dakota plows with eight Belgians and a rope and pulley evener system. This hitching apparatus is demonstrably different from conventional eveners. The four abreast closest to Tom are called the wheelers, The four up front are called the leaders. In a conventional evener system, the four leaders would work together on a four abreast evener that was equalized back to the wheel evener by a lead bar or chain. With the rope and pulley system, each wheeler is hitched in-line to the leader directly in front. There are a lot of ropes and eight pulleys to keep straight (pun intended). If the operator has a problem with something coming undone or tangled inside that hitch, it can be dangerous to unscramble without assistance.. When everything is as it should be, the system works great. Photos by Fuller Sheldon

Willis D. Miller with 12 head at 2003 Ohio Horse Progress Days. Photo by author.

Six abreast of Amish Percherons seen at Horse Progress Days.

At Horse Progress Days many innovations in horsepowered market garden tools have been demonstrated, over recent years. (Left) Creating a raised bed and laying down a plastic mulch sheet.

And different system approaches to transplanting, either direct to soil or through the plastic mulch.

Photos this page by William Castle.

Traffic scene at Oregon Draft Horse Plow Match in Yamhill spring of 2016, Photo by Kristi Gilman Miller.

Horsepower in action for Bernard and Jude Farms in Dexter, Oregon. Photo by Jessica Lindsay.

In many parts of North America, Amish farmers prefer working big hitches all in a line, or abreast. In these scenes from Pennsylvania, a seven abreast and a six abreast do the field work. Obviously, any such units require special arrangements, or very wide gates, when moving from field to field. William Castle of Shropshire England captured these scenes during a trip to attend the PA Horse Progress Days.

Photos from Fuller Sheldon (ND) of a JayHawk stacker. This ingenious device had a short history. It came to use at the same time as tractor innovations and baled hay replaced much of the loose hay operations in the nation. The rolling derrick frame features a top-hung pivoting pyramid frame carrying, at the front end, the hay forks. At the back end of that frame, at the feet of the operator, a cable is fastened which runs forward to a GD drum. When the operator engages the drum the cable is wound as the horses step forward, which raises the basket. I have two of these units and know that they work good on level ground and will raise and dump hay at up to 20 feet.

Ammon Weaver and Athens Industries, of Liberty, KY, builds state of the art treadmills for all different size animals and various applications, from farming to industrial and even domestic use. Above one suitable for the single draft horse, below for a team. With innovations in tread design, clutching and gear ratios, these are brilliant mergers of old power systems to new technologies. Such innovation points to a truly sustainable and exciting future for small scale farming and industry.

Chapter Seventeen

Bits, A Nut on the Bridle and The Wrecking Service

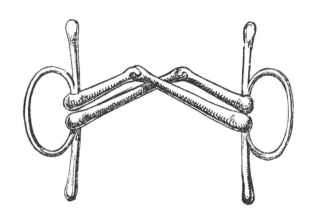

295

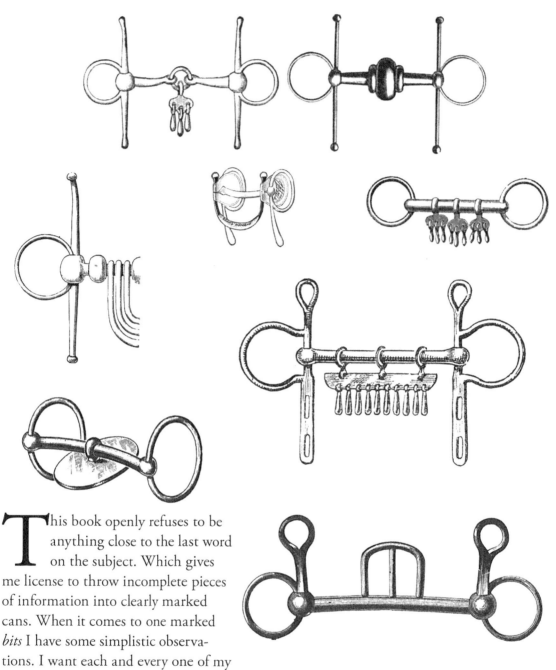

This book openly refuses to be anything close to the last word on the subject. Which gives me license to throw incomplete pieces of information into clearly marked cans. When it comes to one marked *bits* I have some simplistic observations. I want each and every one of my horses to have the greatest possible chance for a comfortable working life. For me that means paying attention to what I ask them to carry around in their mouths. Bits are attached, most of the time, to driving lines that run back to your hands. While the design and functioning reality of the bit may be more or less harsh on the horse, please permit me to suggest that the way you handle the lines is potentially even more so. You need to find a perfect tension, which indeed is actually that point of contact which is just before ANY tension. No slack but

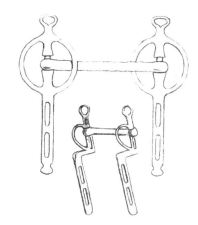
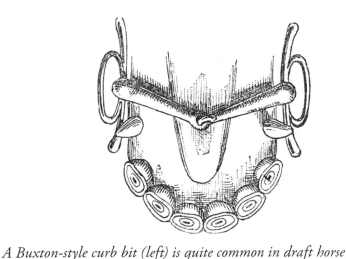

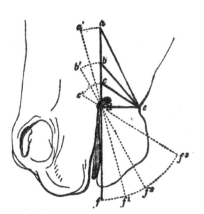

Diagram to show action of curb bit.

A Buxton-style curb bit (left) is quite common in draft horse circles and allows that there be more or less leverage on the curb strap or chain, which runs 'neath the chin. Pull the line back and there is a pinching action on the lower jaw accompanied by a tension on the horse's poll. Compare this to a snaffle or bar bit (below) which transfers all pressure to the bars of the mouth and possibly, depending on design, to the roof of the mouth.

no abuse either. Just almost taut, allowing that very little movement with your hands and arms will give a lot of information to the horse's mouth. Some of the information, if you are aggressive and/or heavy handed, will be painful. Some of the information can be to guide the position of the head for direction and also to gently remind the horse that a slower speed is preferred. If you have soft hands and a sensitive nature, any bit can be made to do the job. If your horse is nervous or comedic you may need to employ a bit that he can't get his tongue over or one which is difficult or impossible to bite down on and render useless.

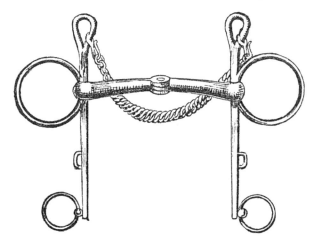

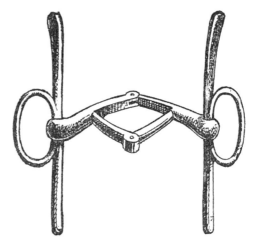

Shouldn't have to say that a bit's length needs to match the width of the horse's mouth. A common issue for folks is having the bridle out of adjustment and allowing the bit to be too loose or too tight. You want to see one slight wrinkle in the corner of the mouth. Too loose, it's uncomfortable and may allow the horse to get its tongue over. Too tight, it's uncomfortable and with time may result in desensitizing the mouth and making the horse less responsive. You want a working horse with five or more forward gears, not one lumbering gear. That starts with good hands and properly fitted bit.

Notice above that one mule is bitted straight away (no leverage on curb chain), and one is bitted at second hole with slight leverage. Using lever bits in this manner allows you to customize the pressure for the animal.

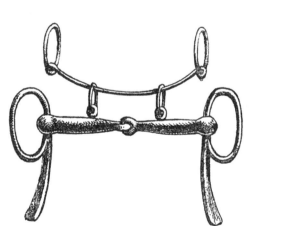
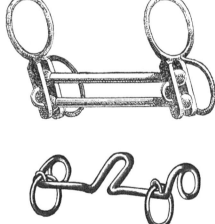

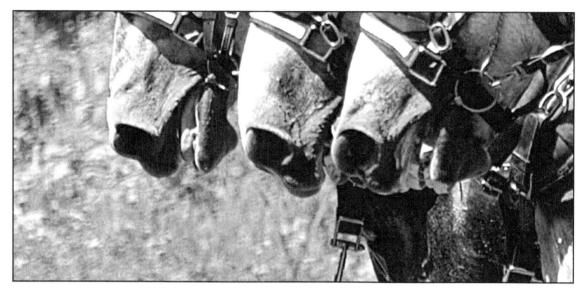

Snaffle bits appear to be at the right adjustment (see that bridle side straps are loose to bit rings). The appearance of tight bits may be from driving lines being drawn back severely by teamster or the result of check reins set so tight that they pull back and up. In any case, compare these bitted mouths with the relaxed ones below to help you recognize comfort.

I have chosen with this volume to avoid any technical information on what I consider to be extreme control devices for horses. I do not believe that such things belong in this book, devoted as it is to the artistry of the craft, something which denotes finesse and subtlety. Things like running Ws, chain bits, buck back straps, and such may have their place, and can certainly come in handy to the good horseman in a difficult situation, but they are typically last resorts. If you need such drastic measures, that information should be sought elsewhere.

It used to make me mad when I was younger but now, in my old age, I understand perfectly that open-ended adage "there are too many good horses…" Meaning if you are struggling to find balance and mutual trust with an animal, move on to a better one.

Here two people are electing to use either a mechanical hackamore, with pressure on the nose, or just a halter. Neither of these are recommended for beginners of any stripe.

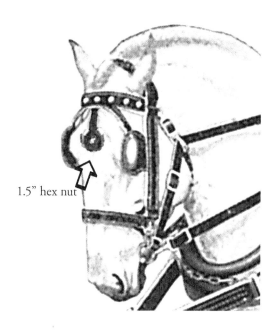

1.5" hex nut

Nut on the Bridle

Back in the early seventies I heard tell of a teamster who was legendary in Oregon in the sixties. As the story goes this man, we'll call him Chuck, could drive any horse or mule no matter how rank. The exact boast he made was "there is no horse alive that I cannot drive". Ray Drongesen knew him and told me this anecdote.

There in the southern Willamette valley another man had a valuable team of registered Percheron mares for show purposes. Whenever he hitched them they ran away. It was determined that the one mare was the culprit. So he brought the pair to Chuck's place. Chuck had the one mare harnessed and hooked to a stone boat in a corral. He wanted to watch the man try to drive her. Sure enough she took off and busted the fence. They caught her (at least she was easy to catch.)

Chuck asked "how much did you pay for that mare?" "$1,000 " was the answer. Chuck said "Give me $100 and leave the team with me for one day. If I cannot drive her I will give you $1,000." The man left the team. Chuck then took the mare's bridle to his shop and slit the brow band dead center and put in a strip of leather to which he affixed a 1.5" hex nut. The nut hung between the mare's eyes when she had the bridle on. He then took the team out to a plowed field where a small foot-lift horsedrawn road grader sat. The blade was down in the loose dirt. He hitched them to the grader and spoke to them. The mare lunged forward, went seven leaps and stopped cold. She shook her head, then held it still. Chuck raised the blade slightly and with light lines, spoke softly to the team and they set off, a little quick, but that mare came a stop again. It took four times, each one slower and quieter, but that mare finally would start calmly and walk along without any indication that it wanted to run.

The man returned the next day and Chuck hitched the team to a wagon and they quietly went through their paces. The man was amazed, then he noticed the big hex nut hanging in the mare's face. "What's this?" Chuck answered, "Whatever you do, never take that nut off her bridle."

The man went up to the mare, team still hitched to the wagon, and said "This is ridiculous, You're not making a fool of me." And he used his pocket knife and cut off the hex nut. Then he went to get on the wagon to drive his team. Before he was in the seat the team took off at a dead run, ending up in a draingage ditch on the edge of Chuck's property. The man went to retreive his team and Chuck went in the house. He came out with a thousand dollars and a rifle. He handed the man the money without saying a word. Walked off ten paces, turned and shot that mare dead. As he walked off he could be heard saying "there is no horse alive I cannot drive."

I asked Ray to explain the story to me. He said, "It was all about the nut. You see when that mare would lunge forward that nut would swing out and then come back and hit her between the eyes. If she walked out calm and smooth the nut would rest there quietly. She was beating herself senseless when running. I guess she had gotten so used to running away, not so much from fear as from knowing that if she ran away she wouldn't have to work, that it was her trick. Just as soon as she knew the nut was gone, she knew she was free to run."

"But did he have to shoot her?"

"Back then, a man's reputation was just about all he had. He set up the wager so that if things didn't work out the man would get his investment back. And for Chuck, the $1,000 was what he was willing to pay to be able to say 'There's no horse alive that I cannot drive'."

Post script: This is one of those real life stories, full of the sort of contradictions and bizarre twists, which narrowly escape being useful. (As several people have remarked "why did he have to shoot her, afterall he did drive her?) This is not a tale of right and wrong. There is no moral to this story. And I am most certainly not recomending any of the actions taken. I share the story because it informed me through the years, helped me to understand the perspective of some of the older teamsters. Allowed me to appreciate how lucky I have been with my horse partners, and how lucky they have been to have me.

The Wrecking Service

My buddy Mike Atkins is a flat-amazing trainer. He loves his Belgians and his mules. And he trains them to accept anything. On lark he trained one of his geldings to stand on a stoneboat and allow his mule to pull him along. He likes to refer to the whole thing as his Wrecking Service. He'll tell stories about having a broke down gelding who needs to be towed back to the farm. And that he has a tow mule who will do it. Can you imagine how many wagers Mike has been able to win with these two at his service?

CAUTION: Boys and girls DO NOT try this at home without Mike there to help

Chapter Eighteen

How Far Back? How Far Forward?

Mexican Beach and an Old Man's Reach

Jacob McIntosh driving the family four-up of Belgians Roman-style. A 'far piece' from "stealing a ride" on the back of an old horse. Photo by Kristi Gilman-Miller.

From whence ...

How far back do any of us go to understand where horses first entered our world, our sphere of influence? It was 1960, I was 13, seven of us in the family drove down in our '55 Buick to a cousin's beach house. Whatever the reason, I was told I had to sleep on the roof, away from everyone else. I remember a flat-topped building with vegetables and herbs growing on the sandy cap of this large rough structure. It was surrounded by grass-capped sand dunes except for the beachfront. I remember smells, the surf sounds, and a blanket of stars so close I was sure some had snuck into my clothing. Though I was being punished, yet again, and separated - I felt blessed to spend those few nights like a bony swizzel stick in a fabulous sensorial cocktail.

Our cousins were Hispanic with no English. I was forbidden from 5 years on, to speak Spanish. As Spanish was my first language, the immersion in rolling, folded chocolate word sounds was like a blanket. I was jealous to witness the happy comfort of my mother as she lobbed Castillian phrases into the waiting air of that short visit, jealous that we never saw that "belonging" comfort in our own home.

Walking on the beach I was fascinated to watch saddle horses being ridden in the surf's shallow back wash. Must have been obvious because a tall cinnamon brown gelding approached me with what looked like a ten year old boy on his back.

After we figured out we had to speak English for my enjoined understanding, he asked if I wanted to ride. He said he would let me ride his horse for a quarter. I told him I didn't know how. With innocence instead of courage I slid up on the bare back of the tall horse. A rope came round the neck and withers to fasten and tie to either side of a braided cord halter. No bit, no hackamore, just a halter. The boy explained by motions how to pull the horse's head in the direction you wanted to turn. He slapped the animal's rump and we walked off along the edge of the receeding surf.

I looked over my shoulder and the boy was gone. Didn't matter. The comfort I felt was real, comfort not security. The horse ambled for a while then entered a bouncing trot. I gripped mane and rope. He made a slow wide arch and trotted away from the ocean and towards the dunes. Then he broke into a long lope. Now I was frightened. I leaned forward, head to one side, and hugged his neck in an effort to stay on.

We hit the grassy cap of a dune and he dug in and lept forward. We were airborn for a very long two seconds and as I looked down I saw two naked terrified adults - one man and one woman - shielding their heads from flying sand as they looked up at the under belly of some great beast - its form distorted by the sun directly above.

We landed hard yet he still picked up speed - then slowed as we approached buildings. There at the open gate stood the little Mexican boy, flipping my/his quarter and whistling for my borrowed charger, my Bucephelus, my first horse.

It would be nearly ten years before I would return to horses and those next experiences accumulated rapidly through ranching procedures, and cuttinghorse ballet, before landing with that first team.

... to where the end was first glimpsed

Bud Diminck was 85 years old and I was 48. He had come out to the ranch with his gelding team and mower to join me in mowing hay ground. He loved the process and didn't have any hayland on his place in Madras, so he invited himself out to help me. We were friends in that guarded way that balances itself when the old master and the young master struggle to conceal how they constantly measure each other.

We were in my ramshackle barn with two tie stalls. My team of mares stood in one double stall and his geldings in the adjoining. I proceeded to harness my horses, a routine that was daily for me and quick. Bud had to carry in his harness from his truck, so it took a little while .... and he was 85.

*Kristi Gilman-Miller mowing with
Cali and Lana, Photo by the author.*

Bud mowing on Singing Horse Ranch, photo by Kristi Gilman-Miller.

Having finished my harnessing and watching him lug in the second harness, I went to grab his first harness to put on the waiting gelding. He dropped his harness and came to me very quick. "Stop that." he said. I pulled away slightly, hanging on to the harness, and said, "it's ok, Bud, I just want to give you a hand with this harness."

With one hand he took a hame and with the other he pushed me away. I tripped and fell. "Don't ever do that again! When the day comes that I cannot harness my own horses I'm done."

Through the tension we got out to the field and started in on a twenty acre land of hay. I opened the land and he followed. Quarter mile along two sides, eighth of a mile along the ends, the first time I got jammed up, instead of stopping his team and waiting for me to clear my bar, Bud drove around and kept going. So, when I was behind him and he jammed up I did likewise. Until I didn't and he flagged me around. I said, "no, I'll wait, the girls could use a blow." meaning I wanted to let them catch their breath. I still remember that faintest smile cross his face. We broke for lunch, went back to the barn, watered and fed the horses, and sat in silence with our sandwiches.

Then back to the field. Round and round, beautiful day. Occasionally passing one another until we had just one thin pass left to do. Bud pulled his team out, tied up his cutter bar and headed to the barn, leaving us to finish that last pass.

When we got to the barn, Bud's team was loaded in the truck, harness still on. I parked the mower and got off to unhook the tugs. Bud came up hand extended and said "Your team is better than mine."

"Oh", I offered, "I'm sure that's not true. Your geldings did a great job today, thank you."

"No, thank you. We needed that work. And I still say your team is better than mine." and with that Bud drove off to his home 40 minutes away.

I got the thick feeling that I had witnessed something private, not meant for me.

The last time I actually saw Bud driving his team he was 90, same geldings, and in a parade. Kristi and I went to his 105th birthday party where he fed himself cake and talked with folks. To me he said, "I always liked hearing you talk." Didn't much matter what he actually said, that he wanted to aknowledge me meant a lot. Kristi took photos that day. Soon after that Bud Dimick passed away in his sleep.
.

Ed Joseph and Lynn Miller mowing on Singing Horse Ranch, photo by Kristi Gilman-Miller

From Kristi Gilman-Miller's photos, the author painted this portrait of Bud at 105.

Chapter Nineteen

How Do You Know When A Horse is Broke?

John's Week and Leonard's Seventy

Distinction: In all my years with work horses, I have noticed a most definite repeated clue to the determining question of equine experience. For the working horse, seeing one of their successful past jobs performed in 'their field' by another team, or a tractor, is achingly fascinating to them. I might have a herd of horses in the adjoining field, and be out mowing with a different team today, and I know that I will see yesterday's team hanging by the fence watching, in a longing way, as we work. And the other horses, ones with no similar field experience, they could care less what we are doing, they are out there grazing.

What constitutes a well-trained or broke horse? And how do you discern that?

John's Week

In the 80's we did workhorse workshops in Ontario, Canada. We had a formula for the many 'away' workshops that I did. This is the routine they followed: the workshops were to be limited to no more than ten students per assisting teamster. Our preference was that the workshop total be either 30 or 40 students, to be determined by the number of available teamsters with horses. In my approach to the organizers I said "We need these teamsters to be experienced and to bring to the workshop their own well-trained animals. These teamsters do not need to be teachers, they do however need to be physically capable and able to defuse any potential problem any student might have with their team. Arrangements were made to compensate these teachers from the paid tuition of the workshops.

In the first year of these workshops we had three teamsters join us; Aden Freeman, Jimmy Grant and John Male. This story is about John Male's contribution that first year. First off let me say that all aspects of the workshop went exceedingly well, even in many cases exceeding expectations.

It was a week long workshop experience with lots of various driving exercises. But for that first day, we used the horses as backdrop as we discussed and demonstrated the nature of the animal; their senses, their conformation, their feeding, and physical care including hoofcare,

The best teachers are the well-trained horses themselves. Here's Jenny the Fjord teaching Scout Miller. Photo by Kristi Gilman-Miller.

*King and Ruby mowing with Ray Drongesen on author's
Junction City Farm, 1975. Photo by Nancy Roberts*

the mechanical aspects of how they translated forward motion to work including harness and harnessing. The very first thing I tried to establish is empathy from the students. I wanted them to understand how they might be viewed from the horse's perspective. And the best way to do that was to speak to and about what was actually happening with this workshop.

These horses did not know us, only their own respective teamsters. Here came a crowd of folks with notepads and clipboards, staring at them, poking at them, moving them around. For example, I had the students standing in the middle of an enclosed arena building and the teams standing, tied outside of that group of people. I asked the students to observe how the horses reacted when we stared at them and walked with phony, hesitating caution towards them. "Watch their ears, eyes, nostrils, head-set for signs of nervousness. Now all of you avert your eyes, talk to one another in a matter-of-fact tone and walk casually away from the horses. From the corner of your eyes what do you see different in the posture of the horses.?"

We did these sorts of inquiry exercises, over and over again, with each horse. We were doing it to teach the students and hopefully have them develop a real empathy for the nature of the beast. Then we moved on to picking up the feet. I wanted to show them where to stand and what to do if a horse was less than willing to give a foot. Working around I came to one of

Lise Hubbe of Scio, Oregon, driving her four-up on riding plow. Photo by Dick Brown.

the roan mares belonging to John Male, we'll call her Ginger. She was unknown to me as were all of the horses. I noticed she was nervous, and John seemed a little nervous as well. When I went to the front leg he stopped me and started to explain something. I interupted him. "I appreciate that you want to caution me here, John. But please don't. What I want is for the students to see what I might do to safely determine how she will react when I pick up her foot, and if she won't let me take it, what I might do next. I want to approach her knowing nothing, which is most likely what may happen with the students later in their experiences."

I stood where I needed to be and ran my hand down her front leg, she immediately pulled it away and stomped. So, explaining myself to the watching students, I ran my hand down her leg slowly, and when I felt her tighten I slowly pulled my hand up a little, doing this over and over again until I could run my hand all the way down without her reaction. I did NOT try to pick up the foot. Next, I backed her a step or two to where her weight was off that leg and, with no caution, I smoothly reached for her pastern and lifted the foot and released it BEFORE she could take it away. I repeated this two more times. Then we left her alone and proceeded to another horse. (I told the students we would return to the anxious mare in a few minutes.) The second horse was quiet and willing for the exercise.

A few minutes later we returned to the mare and I asked a student to repeat the exercises I had demonstrated, while John Male, Ginger's owner attended. I and the rest of the class went on to another horse. I rotated students back to that first mare to repeat the gentling with her front legs. Before lunch break I went to the mare and picked up the foot and cleaned it with a hoof pick and gave it back to her. None of this was done for the mare's sake. It was all done to demonstrate to the students an approach to the situation. Turns out this distinction was vitally important.

And so the workshop proceeded; three assisting teamsters with three different teams. Jimmy Grant worked a stallion and a mare, Aden Freeman worked his championship Belgian gelding team and John Male had his two strawberry roan mares. Whenever any horse demonstrated any hesitation with the exercises at hand we used the opportunity to teach the students how that might be handled, but that was secondary to a process that taught them how to drive horses for work. This meant that over the week we drug sleds, pulled wagons, skidded logs, ran cultivators and plows, and always with great care for safety. Believe me when I say you cannot imagine all the ways nervous students, lines in hand for the first time, might befuddle themselves and the situation. For the students it was a wide range of new experiences.

At the conclusion of the workshop, all of us together, John Male wanted to "come clean". This is the story he told us.

"I was excited when they asked me to help with this workshop. I also needed the money. So, when I went out to the barnlot and found that one of my team was bad lame I got busy trying to find a replacement. At the stockyards that week I bought Ginger, brought her home and found out she didn't know much and was a nervous handful. I probably should have told Lynn but, like I say, I needed the money. So I hoped I could work out her bugs and get her safe. When we started here on Monday she was still as green as they come. And I was

John Male and Ginger nearest us. Photo by Kristi Gilman-Miller

too frightened to say anything. Then we all watched this miracle, only you folks didn't know what you were seeing. From the beginning with her feet, and all sorts of different people touching her, then to harnessing and unharnessing over and over again (by the way she didn't like being harnessed before we came here) to ground driving and driving single and double to skidding and even plowing, she got better every single step of the way, until - as all of you have seen - she became the quietest gentlest most willing horse here. What Lynn did was a miracle. He trained this difficult horse right in front of us."

I was listening to John intently and I could see how the pattern unfolded. I said to them all,

"Thank you, but it wasn't I who transformed Ginger, it was the class and the process. Coincidentally we subjected her, in an intense week, to as complete and wide ranging a set of new experiences imaginable - and we gave her, indirectly, only one option, to improve with every step. But, as I see it, one important piece of this puzzle is that, at no point, were we addressing HER as a problem that needed correcting. WE, each of us, came to her with the mindset that WE were learning. As I have told you throughout this workshop, I believe completely

Photo by Jerry Hunter at 2016 Horse Progress Days. Sometimes the training level and disposition of a good horse can be read with a glance.

that the equine has a telepathic sense we can only guess at. I submit that you have just witnessed a dramatic example of that. She read us, telepathically, and we sent her no scary signals. No one here did anything to mess with her mind. We did not attempt to crawl up inside of her head, to psyche her out. We joined her, in her space, and we learned as she did."

Leonard's Seventy

Way back in the eighties we attended Small Farm Gatherings in Missouri. At the first one, amongst a gaggle of exceptional teamsters we met Leonard Mothersbaugh, a big, opinionated farmer who drove a four abreast of three year old Blond Belgians, well matched and exceedingly well-trained. For a long day I admired them at every turn. During a round table discussion on training workhorses all sorts of different schemes, approaches, devices, and plans were revealed. Some of what I heard made me squirm. But my place was as moderator so I kept my tongue. Truth be told, Leonard at one point interupted the conversation and made a

Stephen Leslie of Vermont, cultivating with a Fjord.

Walt Bernard running a cultimulcher in his hoop house. An indication of the sorts of working environments that well-trained horses will accept.

lame proclamation as to how women should not be allowed to handle stallions - period - and equated them that do with the wrong side of the right to life equation. Tense moment. I attempted to turn it away and change the subject back by asking him straight away,

"What do you do, Leonard, to train your horses?"

"Magic number," he said "is seventy. All you got's to do is hitch and unhitch them to something, whatever it is, seventy times. Don't care if you just go out and drag something for fifteen minutes, so long as you do it seventy times, and without no mishaps. Once you hit that magic number they are broke. I 'garoantee' it."

"So, how about that four you're driving today?"

"I stop counting at seventy but if I had to guess I'd say they've been hitched at least twice that. Hell, they are so broke now I'm 'bout tired of them."

Made me think back to John Male's mare, Ginger, in the workshop. Thirty students doing two things with her each day for five days that comes to.... 300!

Above: a hayload on Littlefield farm. Below: a new Italian manufactured hay rake.

Once had a man ask of my horse, "where will it go without fear, and how long can I leave it unattended in the field?" Photo in Horsepower Vineyard, Milton Freewater, Oregon, from Joel Sokoloff. Below, photo from Eric Nordell of a patient good one.

Over the years I have had many grape and cane fruit growers tell me that they wished they could use work horses for cultivation but just couldn't see them working in that tight and restrictive environment without doing considerable damage. The good, quiet, attentive horse, I submit, will do less damage than most average human farm help. Case in point; looking at this picture from Horsepower Vineyard, Milton-Freewater, Oregon, of a Belgian cultivating grapes, ask yourself "what does the teamster see?" Nothing? That good horse is watching for him. Photo from Joel Sokoloff.

Pictures by Don Longmire of Aaron Ebersol's all weather plow cab, constructed to put over the top of a Pioneer riding plow for cold-wet weather. Some of us have known horses which when asked to hitch to this contraption most certainly would have thrown a fit. While others, with the right training and conditioning, just might accept it as another day at the office. This unit has a battery-powered light mounted on top for dark, early morning work.

Chapter Twenty

How Deep It All Goes

Bud and Dick

Again I say, I believe that horses have a telepathic capacity. Stand behind a team of harnessed horses with blindered bridles and concentrate on one of them. Look at the back of that horse's head and just concentrate. Before too long, you'll notice a reaction. The ears will move, the head will move. The horse is acknowledging that he is sensing something. He's sensing a communication from you, a wordless communication, communication without touch.

This was brought home to me in an experience I had back in 1975, one of those experiences that has gone on to shape my entire sense of and time with working horses.

Bud and Dick were a pair of beautiful, bay, Shire-Percheron cross geldings. Together we had spent a lot of time working in the field. The spring in question, my good friend Ray Drongesen had loaned me his Oliver 23 two-way riding plow, the one that he had modified by taking one bottom off of it. He made this change so that the plow would work better for contest purposes. He and I were not only trying to get our spring work done but we were also tuning up our tools and skills for an upcoming state plowing match. I had a specific goal, I was anxious to do a better job of finishing up the dead furrow and this particular plow was exceptional for that purpose. Exceptional because it had a single axle, two wheels, with the one bottom. You didn't have to worry about the articulated steering that you would normally get from a foot lift riding plow, one that has a 'steering' tongue that will turn the front and back wheels. With those models of plow, any variation at all on the end of that tongue would cause the plow to steer one inch or two inches either way and prevent you from doing a nice clean last pass on that dead furrow.

Anyway, before Ray had lent me the plow he warned me. He said, "I put that big washer on that seat because I think that seat is weak. I think there's a lot of rust underneath it. You better check that before you take it to the field. And also, it wouldn't hurt if you took that rolling coulter off and sharpened it up."

When I got the plow unloaded out of the back of the pickup there at my farm in Junction City, I kind of wiggled that seat and it felt awfully solid to me. Plus, sure enough it had a very big washer on the bolt head. So I figured it would be okay. I took the rolling coulter off and sharpened it till it was just like a razor, and put it back on and put that good pair of geldings, Bud and Dick, on and went to plowing.

And I was coming up to finish out that last dead furrow pass. The horses had been doing this work for me for so long that they were making some decisions for themselves. It was pretty much second nature. Walking in the dead furrow for that long pass, I wasn't concerned about driving them, so I tied a little loop and hung the lines on one of the handles right in front of me. This allowed me both hands to adjust the two handles to the variation coming through the wheels. Each wheel was riding on the tops of the plowed furrows, up and down and sometimes sideways. It was a challenge to keep the plow level but I had to try otherwise that long, narrow, last strip of sod wouldn't flip over as it should. I'd have Mohican strips left over and that would have docked me in the judging.

The dead furrow strip.

The horses walked straight and we were doing a real pretty job of cleaning up that last dead furrow strip, but that seat was rocking back and forth, and back and forth, and all of a sudden the seat broke off and I found myself grabbing forward for those handles trying not to grab onto the lines, … grabbing forward for those handles, trying to stand up on the axle of that plow. Slow as we went, it still felt like things were moving too fast. I muttered and sputtered a whole succession of "whoas" and the horses stopped right on a dime, even before I quit sputtering.

I found myself rolling forward. I avoided the lines, grabbed hold of the handles and then the plow tongue and found myself falling, awkwardly forward, tangling in the evener with my legs coming underneath the tongue. With the last little forward motion of that plow I could feel the bite of that sharpened coulter in against my one leg.

I laid there under that plow tongue. It took me half a second to realize that if those horses stepped ahead, even a half a step, I was going to lose my leg at the thigh. I tried to regain my presence of mind. I concentrated as hard as I could. I didn't say anything for fear that they would take the sound of my voice to mean that I wanted them to step ahead, I didn't say anything at all.

I concentrated and I thought, "Oh, please, dear God, if you do anything, Bud, Dick, if you do anything at all, please just back up. Just back up." and I concentrated and looked up at them.

As I was laying there, off balance and tangled up, thinking these thoughts, I noticed that Dick, the left hand horse, turned his head in towards the tongue. He looked back down the middle at me and then he looked straight ahead. Then Bud did the same thing, he turned his head towards the left and looked down that tongue, looked at me. Then he rolled his head around all the way to the right and looked back down along the outside, taking measure of the situation. I kept thinking to myself, "please, if you do anything at all, please back up". And I'll never forget what happened next.

In unison, in perfect unison, these two horses set their feet and leaned back and took together one step backwards freeing my leg. I very carefully, slowly, slid out from under there because I didn't want my motion to maybe cause them to step ahead. When I was out from under there, I walked around in front of those horses, gave them both a big hug and kissed them.

Then I went back to try to figure out what had happened. I went to the back of the plow and I could see there had been a lot of rust, a lot of cracks, and a lot of crystallization, Ray had been right, that seat had just popped off. I should have listened to him. I should have replaced it.

The job needed to be finished so I walked behind the plow and drove the boys from there. We finished out that last little stretch of the dead furrow strip. I was so proud of those horses, what they had done. When I got back to the barn, I took the horses into their stalls and fed them, pulled the harness off, and curried them down. We were done for the day. But I was feeling like I owed them a treat.

What I did next… I don't know if I can ever forgive myself. There was a little, small quarter acre patch near the barn where the rye grass was growing thick and luxuriant, about knee high. I wanted to give Bud and Dick a treat, a thank you present for what they had done for me. So after I got the harness off of them, I took them around and turned them into that patch and gave them that grass to eat. I figured I would be careful about it, come back and pull them out after fifteen minutes, to a half an hour, because I didn't want them to get too much and get sick. So I went in the house.

I came back out in fifteen minutes and got the two boys out of that patch and took them back into the dry lot near the barn where I kept them during the working season. I noticed that Bud just didn't seem to be feeling very good, he was acting a little odd, and I watched him. He seemed ok, just slightly uncomfortable. It had been a long day's work. I went back to the house. Something was nagging at me so I went back out about an hour later and he was walking around in circles and pointing his nose towards his belly. I knew those were tell tale signs of colic and I called a vet, an excellent vet, not far away.

He showed up within the half hour. He gave the horse some muscle relaxant shots and gave him a drench of mineral oil, and he gave me strict instructions on what I was to do. I had to keep walking him, not let him lay down.

I spent the entire night with Bud, talking with him, apologizing and worrying as I saw him go deeper and deeper into pain. In the deep darkness just before sunrise he went down.

Crying I got down with him, and with his big head in my lap he died.

It goes that deep.

Chapter Twenty-one

Fine Tuning

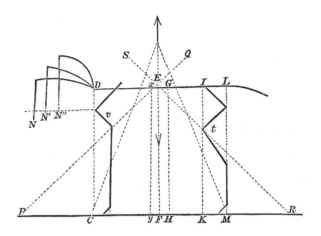

H ere we are at the end of the book. Me, I can't see it. The book I mean. Been working on it for so long that it's invisible to me. Don't know if it is any good. But then, I'm less concerned with that than I am whether or not it may *do* any good. And on that point I feel somewhat assured.

Because I am able to back away from the process and reflect instead on this day and age we live in. Yesterday I skimmed over a scientific article which made observations about our short history with genetic engineering and how the grand promise had failed; and how the horrible price to pay for the exercise is still being computed. I am interested in these observations now because they point squarely at terrible questions of the age, swirling around all things digital and how it is that we have pretty much traded away living artistry for these dead calibrations of virtual reality.

Man's long history has been one of hand offs. The Summerians and Phoenecians handed off to the Egyptians and Greeks and they to the Romans and Visigoths who somehow threw long balls back and forth with the Chinese and the Incans and the Harvard graduates. We humans grew as a species because of those successful and terrible handoffs. But there had to BE something to handoff. Language to moveable type, plant collection to plant breeding, mold to medicine, sunlight to electricity. Now we are handing off hard drives? There is no there there.

I have used the analogy before: I see collecting and saving archaic technologies, crafts, and arts as a form of seed saving. The art and craft of working horses as a power source I see as a

CLYDESDALE STALLION: LORD STEWART

parallel to glass blowing or to steam engines or to the firing of clay pots or to natural farming. These all contain procedural truths visible only to practioners, only to people deep in the process. If we save seeds in a jar, never touching them, never using them, their vitality and their secret powers rebirth will be lost to us. We have to plant the seeds, steward the plants and collect the next generation of seeds, repeating the cycle. The cycle must be repeated or the secrets are lost.

Because of its many attractions, the art of working horses has remained attractive to enough people to keep the cycle in motion, the seed alive. And one step further, the growth in interest begs the quasi-rhetorical question; "but is it a viable way forward for the future?" Flipping through this book you see a cavalcade of pictures which give clear evidence to what working horses are doing today. You see clear evidence of the wide variety of people doing it. And, hopefully, you see the light it generates giving the clearest view of the secrets and magic it is shaped by, and holds. It is my hope that you also see that this wonderful marriage of humans and equine does offer one way of getting work done in supreme balance with nature.

'Fine Tuning?' Yes, calibrating something to a standard frequency. In this case allowing that the art of working horses move usefully into the our future. In this case the *frequency*

SKIN MARKINGS—II

Blue Roan, White Stripe   Gray
            Piebald
Red Roan, White Stripe   Cream

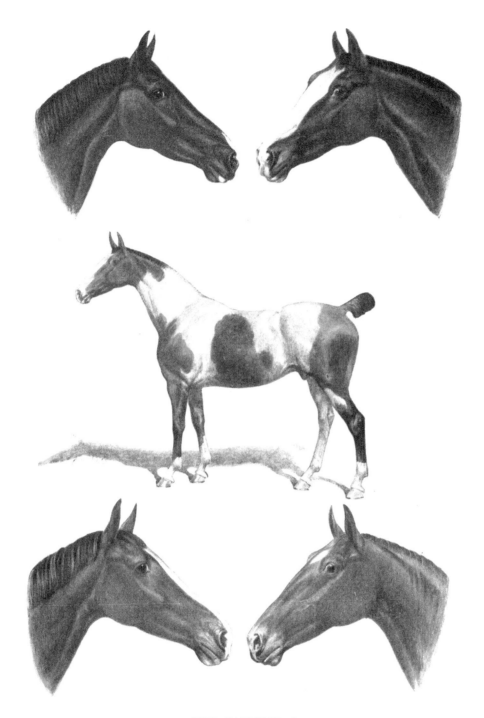

SKIN MARKINGS I

Black, White Snip on Nose   Brown, White Face
Skewbald
Bay, White Blaze   Chestnut, White Star

Jiggs Kinney and nine head plowing Iowa.

is viability in our time. It may require finesse to skillfully answer the rhetorical questions of obsolescemce but only if that is where we want this discussion to go. I prefer to talk about all the incredible opportunities there are out there for applying this power in attractive, sustainable, and forgivable ways.

Today people tell me that computers are just tools which allow us to do more. But I cannot see them without seeing behind them the motives of powerful shortsighted people whose goals appear to be for more power (over us) and more short-sightedness (for them). As the world embraces the rush to legalize mood enhancing drugs and coin-operated escape from living reality, I believe more and more people will come to artistic and craft-based process with a passion to reconnect their own lives to rewarding work and balance.

Photo by Jerry Hunter taken at the 2016 Horse Progress Days.

So I leave you with more pictures to make the case, while answering 'that' question. Yes, emphatically YES, the art of working horses has a future. May I add that also, along with many other artforms, it IS the future.

Steve Hagen cultivating Oregon vineyard. Photo by Harley Hagen.

In three photos William Castle demonstrates a couple of fine tunings for single horse plowing. Above: ploughing in a cover crop of rye, the furrow being about 7" wide and 5" deep. The draught chain is well over to the right because the rye crop provides very little 'bite' to the plough.

Shallow 2" skim plowing illustrating an offset hitch.

The draught chain is more central when ploughing sod, as even when only ploughing shallow, the plough holds itself in line.

Comfort Zone

* Horses respond to and are shaped by comfort and repetition
    * Knowing what to expect gives comfort to horses
* Comfort equates with kindness
    * Comfort can be aggravated by silly coddling
* Tickling and pestering and general sugared sappiness can strip comfort
    * Sometimes comfort means space and being left alone

Paying close attention to a horse, intense eye contact, constant small talk, whining and wheedling around them, will make some individual animals nervous and uncomfortable

This photo by Kristi Gilman-Miller was taken in 1980, up in the Okanogan highlands of Washington. Gary Eagle is discing his field. Gary, a blacksmith and artist, is a legendary horseman who inspired many by his dramatic example.

This Friesian horse is cultivating Jelmer Albada's asparagus patch in Holland with a traditional Dutch "zig zag" harrow. Next page; same horse different cultivator.

Mike Downs plowing with his Belgian team at Kentucky Spring Field Day on his farm. Photo by Rick Conley.

Eric Nordell rigged a roller gang on his straddle row cultivator for special market garden application

2 stallions and a gelding hitched in a 'random' (or in-line-three). This breed is called 'Trait Belge' in Germany. Horses belong to Patrice Breuskin, Claudy Lux, and Bruno Huybens and were hitched twice for this logging demo, at Pferdestarke in Detmold, Germany. Photos by William Castle.

Dick Brown discing with 8 head and JD 13' tandem disc he switched over for horses using battery-powered pump to raise and lower. Homemade tongue truck up front.

Tom Odegaard cultivating corn in North Dakota. Photo by Fuller Sheldon

Farming in Norway with Norwegian Fjords.

Grinding corn (above) and planting grain (below).

Photo by Bernard Chambers, shared by William Castle, of Jonathon Waterer with Shires drilling seed in North Devon, England

Jenny Golding's Fjord in Norway

Photograph by Richard B. Hicks of Tim Reynolds from Clarissa, MN with Belgians Bonnie and Trouble and a Percheron/Morgan cross Trace in the furrow. He's using a New Deere Plow

Ken Geiss with his miniature team on a custom forecart and mini spreader.

William Castle's spectaular picture of a Pennsylvania Amish mule hitch headed to field trials at Horse Progress Days. Some Amish prefer mules (especially in PA) or Belgians, or Percherons. A few like Painted Draft and Clydes or Shires. Many prefer Haflingers. The best breed of work animal is the one you never tire of looking at.

Ian Bannerman of New South Wales, Australian training Reuben. Photo by Sandra Bannerman

Photo from Pferdstarke by William Castle. Rollover plow.

Mick Massey, sent us these pictures of his Shires working in Norfolk England.

(Below) Herbert with a load of hay

(Left bottom) Logging with New and Moses

(Below) Herbert and trolley at horseshow.

Hugo Sanheuza of Colombia with his Brabant horses, custom-built forecart and sod aerator attachment.

Hermand drives his Ardenner gelding through a log demonstration course at Detmold, Germany. Photo by William Castle

(Above; Ryan Foxley plowing on Littlefield Farm. Photo by Joe D. Finnerty

Candy drives the Hegele Clydesdales in Oregon. Photo by Kristi Gilman-Miller

A closeup view of the appartus involved in the rope and pulley system, with three spans of four abreast mules at Pennsylvania Horse Progress Days

Ray Drongesen in 1977, driving Bob and Bud on three wheeled forecart designed by the author.

Jess Ross driving his Percheron team on a buckrak.. Photo by Kristi Gilman-Miller

Lynn Miller driving McIntosh binder in Terrebonne, Oregon. Photo Kristi Gilman-Miller

John Erskine of Washington, plowing with six Shires. Photo by Heather Erskine

Bub Roux, Beavercreek, OR, driving his mares, Sally & Floree, in the walking plow competition. Photo by Miscka Press

Chad Loxtercamp from Coon Rapids, MN with Shires Heather and Shadow on a Minnesota one row cultivator. Photo by Richard B. Hicks

Jeb Michaelson, photo by Kristi Gilman-Miller

Barry Gorsenger from Mora, MN with his gray Percheron mare May on a McCormick Little 4 one horse mower. It has a three foot sickle/bar. Photo by Richard B. Hicks

Marvin Brisk of Halfway, Oregon

Homemade Manitoba round bale mover.

Steve Hagen, Old School Vineyards, Oregon. Photo by Harley Hagen.

Horsepower Vineyard, Milton-Freewater, Oregon

Chandler Briggs, of Oregon, cultivating the market garden.

Back in 2003 Small Farmer's Journal subscriber Mary Bolin of Anchorage, Alaska, sent us these two remarkable photos her husband took. He looked out the window one sunny morning and the horses were set as you see them above. He grabed his camera and caught this picture. Then..

He whistled and, without moving a step all of them turned, with ears up, and looked at him as he caught this second picture.

Reminds me that, just as with the stories in this book, you can't make this stuff up.

John Dodd ploughing land before sowing turnips on Silly Wrea Farm, Umberland, England farm with Clydesdales, Photo by William Castle

About the author:

Lynn R. Miller is a farmer, a horseman, an artist, and a struggling writer. He and his wife live on a remote, old-fashioned ranch somewhere on the east side of the Cascade mountains. He has farmed organically for fifty years; worked horses for almost that long. He has been a painter for all of his life. And he's still trying to be a writer. He is currently working on a novel, a book about painting, a poetry book and a book about horsedrawn drills and planters.

Some people see him as old. He prefers to think of himself as a work in progress

Lynn R. Miller. Photo by Kristi Gilman-Miller

CPSIA information can be obtained
at www.ICGtesting.com
Printed in the USA
BVOW05s1156181217
503110BV00021B/1172/P

9 781885 210180